John Lea

Tourism and Development in the Third World

London and New York

To Dee

First published in 1988 by
Routledge
11 New Fetter Lane, London EC4P 4EE
29 West 35th Street, New York, NY 10001

Reprinted 1993, 1998, 2001

Routledge is an imprint of the Taylor & Francis Group

© 1988 John Lea

Typeset by Hope Services
Printed in Great Britain by
Clays Ltd, St Ives plc

British Library Cataloguing in Publication Data
Lea, John
   Tourism and development in the Third
   World. — (Routledge introductions to
   development).
   1. Developing countries. Economic
   conditions. Effects of tourism
   I. Title
   338.4'791091724

Library of Congress Cataloguing in Publication Data
Lea, John P.
   Tourism and development in the Third World / John Lea.
      p.   cm.—(Routledge introductions to development)
   Bibliography: p.
   Includes index.
   1. Tourist trade—developing countries. I. Title. II. Series.
G155.A1L39 1988
380.1'4591009172—dc19

ISBN 0–415–00671–6

Routledge Introductions to Development

Series Editors:
John Bale and David W. Smith

DPV
(Lea)

Tourism and Development in the Third World

For many Third World countries international tourism is the mainstay of the economy but political instability or an environmental disaster can divert the flow of visitors for years. Many smaller nations see tourism as a last resort rather than a development priority, allowing the industry to slip increasingly into foreign control.

This introductory text provides a critical appraisal of the approaches used to understand and categorize the tourist industry. Up-to-date case studies illustrate the critical interplay of diverse social, economic, and environmental impacts, and the diversity of problems facing the host communities. The author explores ways of managing tourism as a resource and evaluates its long-term contribution towards national development.

John Lea is Senior Lecturer in the Department of Urban and Regional Planning, University of Sydney.

In the same series

Contents

Acknowledgements

The author would like to thank the following for kind permission to reproduce material for this book: Euromonitor Publications for figure 1.1 and table 3.2; United Nations Environment Program, Nairobi, for figure 1.3; the National Centre for Development Studies of the Australian National University for figures 2.1 and 2.2; A. Mathieson and G. Wall and Longman for figure 2.3 and much inspiration besides from their book; Butterworth Scientific, P. Wellings, and J. Crush for figure A.1; Visa Travel Agency, Sydney, for plate B.1; Bruce Miller and the *Sydney Morning Herald* for plate B.2; W. R. Henry for figures D.1, D.2, and D.3; Adventure World and United Touring Company for plate D.1; *New Internationalist* Magazine, PO Box 82, Fitzroy, Victoria 3065, Australia, for figure 5.3 and plate E.1; and Pluto Projects/Pan Books for figure E.1. Thanks are also gratefully due to John Roberts and Denis Jeffery for preparing the illustrations and to Jennie Small, my research assistant, for many hours spent in libraries and on the telephone.

1
General introduction

International tourism has been described by Louis Turner as 'the most promising, complex and under-studied industry impinging on the Third World'. It is only recently, however, that tourism has begun to take its place besides more traditional economic activities in the textbooks on Third World development. When one examines the large and rapidly growing literature on tourism, some interesting trends and divisions emerge. Up to twenty years ago all studies tended to assume that the extension of the industry in the Third World was a good thing, though it was acknowledged that there were a number of associated problems to be overcome in time. This picture changed in the 1970s with academic observers taking a much more negative view of tourism's consequences to the point of forthright criticism of the industry as an effective contributor towards development.

A peculiar pattern of rich, temperate, countries of origin connects tourists to a much larger number of less affluent and warmer destinations comprising a 'pleasure periphery' on a world scale. This has been identified as a band of host countries stretching from Mexico and the Caribbean to the Mediterranean; from East Africa via the Indian Ocean and South-east Asia to the Pacific Islands; thence back to Southern California and Mexico. Most of the tourism in this periphery is actually confined to its most developed parts such as Singapore and Hawaii, emphasizing the fact that the industry, in common with other major sections of world trade, flourishes best in the more economically advanced countries. It should also be noted that most tourist movements are not international at all and are confined to domestic travel in the rich western nations.

Another way of illustrating international tourism is to portray the world as eight geographic regions (figure 1.1). Regional growth in tourist arrivals is shown in the top map as totals of visitors for the years 1975 and 1980, confirming that the largest markets are western and eastern Europe and north-central America. The picture changes in the bottom map where the rate of regional population and tourism growth is measured for the same period. Here we see that the largely peripheral regions of the Far East, Oceania, and Latin America are outstripping the rest in the pace of tourism development.

In terms of sheer size international tourism accounts for about 6 per cent of world trade and more than 300 million tourist visits annually. One projection expects visitor travel to rise to 500 million by the year 2000. As far as individual developing countries are concerned some, such as the small islands in the Caribbean, are highly dependent on foreign tourism for export income, whereas large states like Brazil or India are much less vulnerable and more diverse in their earning capacity. Others, like Tunisia and Sri Lanka, have experienced a roller-coaster effect as the rapid boom in tourism of the 1970s has been replaced in recent years by falling visitor numbers and increased competition from alternative destinations.

International tourism is a volatile industry with potential visitors quick to abandon formerly popular destinations, like Beirut or even Greece, because of threats to health or security. It will be clear by now that a range of pressures and peculiar influences affect tourism's role in development and distinguish it from the conditions governing the agricultural and mineral exports making up the bulk of Third World trade. Of special significance here is tourism's classification as an invisible export industry, along with banking and insurance where there is no tangible product, and the fact that tourists have to travel to a destination to make personal use of the facilities they desire. This means that few freight costs are usually incurred by the Third World host country because the means of travel is generally foreign-owned. In addition there are obviously different kinds of face-to-face contact between the visitor and host society.

There is no other international trading activity which involves such critical interplay among economic, political, environmental, and social elements as tourism and this is demonstrated in the large and rapidly growing literature. The temporary presence of mainly white and relatively wealthy foreigners in tourist hotels, situated at scenic locations in some of the world's poorest countries, is a scenario fraught with risk and opportunity. In this book we will investigate how tourism contributes to development in the Third World and outline some of the approaches used to understand and categorize the industry as a distinct form of economic activity. Case studies selected from the major tourist regions will draw together some of the more important lessons from current experience.

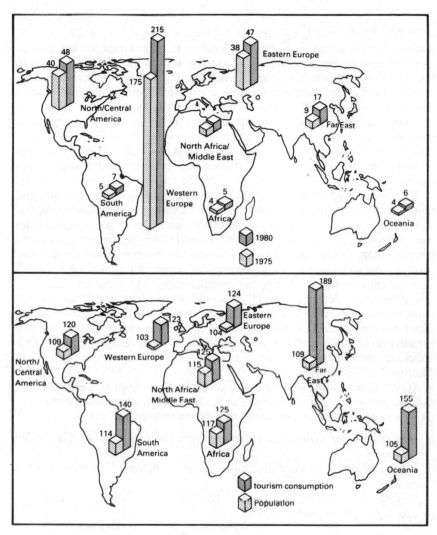

Figure 1.1   (top) Regional growth in tourist arrivals 1975 and 1980 (millions of visitors); (bottom) Growth in regional population and tourism consumption 1975–80 (1975 = 100)

Data Source: Senior (1982: 70–3)

Some definitions

Labels like 'tourism', the 'Third World', and 'development' are part of popular terminology and capable of many different meanings according to the purposes for which they are used. The word tourism for example is capable of diverse interpretation with one survey of eighty different studies finding forty-three definitions for the terms traveller, tourist, and visitor. Perhaps the most useful and comprehensive definition for our purposes is the one suggested by Erik Cohen who describes the tourist as 'a voluntary, temporary traveller, travelling in the expectation of pleasure from the novelty and change experienced on a relatively long and non-recurrent round-trip'.

The term 'Third World' is equally capable of confusion and is meaningful only when contrasted with industrialized nations (the so-called First and Second Worlds) of the western democracies and eastern socialist countries (see John Cole's book in this series). We will follow the approach used by Stephen Britton who uses the terms 'centre' or 'metropolitan' for the core western democracies of North America, Europe, Japan, and Australasia but excludes the Eastern European socialist states. The remaining countries are collectively described as 'Third World', 'periphery', and 'developing' with the general exception of mainland China, Cuba, and Vietnam which have adopted socialist forms of government. The reason for leaving the socialist countries out of the general definitions will be apparent when we examine theoretical approaches towards tourism and development in chapter 2.

Development is the most slippery concept of them all and is commonly used in many different ways as well as having changed in meaning over time. John Friedmann has identified at least five dimensions:

1   To suggest an *evolutionary process* of a positive kind such as rising incomes.
2   In association with words like under, over, or balanced, suggesting it has a *structure*.
3   As the development *of something* like a society, a nation, or skill.
4   As a *process of change*.
5   As the *rate of change* at which the processes occur over time.

In the past the measurement of Third World development was limited to certain economic indicators such as per capita income (averaged over the whole population) and gross national product (GNP). This said nothing about how wealth was distributed, nor did it identify the status of important social factors like health, education, and housing. Today development is seen in much broader terms recognizing that increases in national wealth may benefit only a small élite and can, in extreme cases, hide a decline in living standards for much of the population. Some international agencies, such as the World Bank, regularly publish comparative social indicators covering many useful criteria which can be used in conjunction with the more traditional economic measures to provide a more balanced picture.

In this book we will consider the contribution of tourism to Third World development from important historical, economic, political, social, and environmental perspectives in recognition that economic benefits from the industry in the host countries can be outweighed by less obvious disbenefits in other areas. It is open to question, for example, whether locating luxury casinos in very small Third World countries, with every possibility of associated crime and prostitution, can be judged as development from the viewpoint of most of the population. Similarly, spending huge sums on a new international airport may put a small island on the world tourism map but has every possibility of reducing essential investment in other areas and making it easier for skilled islanders to leave their homes for higher incomes elsewhere. The answers to these dilemmas are rarely clearcut and have led to a serious reappraisal of international tourism in much of the Third World.

Primary elements of international tourism

It is very likely that the thoughts which most of us have about tourism are centred on the travellers themselves or on a variety of exotic destinations, without considering who actually controls the international market. We take for granted today the existence of travel agents and tour, hotel, and transport companies which intervene between the would-be tourist and a chosen destination. These intermediary companies created and now control a global mass tourism market through their transnational operations in origin and destination countries. The largest have expanded vertically to the point where they have interests in all major sections of the industry and are not answerable to policies determined by any single government.

This evolution of tourism into a business organized along the lines of transnational manufacturing industry has brought with it specialized marketing techniques like the packaged tour, fly-now pay-later deals, and the powerful persuasion of radio, television, and the print media. The modern tourism product is not confined to travel and accommodation but includes a large array of servicing activities ranging from insurance to entertainment and shopping. Even the creation of the desire to travel is covered and, we will demonstrate, is responsible for the emergence of a travel mythology which often bears little resemblence to reality.

Figure 1.2 summarizes the major influences on both ends of the tourism experience and emphasizes the strategic position of the intermediary companies. There will be an opportunity to examine some of these items in much greater detail later but they are listed here to give an early idea of the threefold division involved. Certain company groups like airlines and banks have used their strategic positions to expand into many sections of the industry, minimizing the functional divisions between activities like marketing, accommodation, and travel. Differing explanations of how this arrangement works in practice and its effects on Third World destinations will be examined in chapter 2.

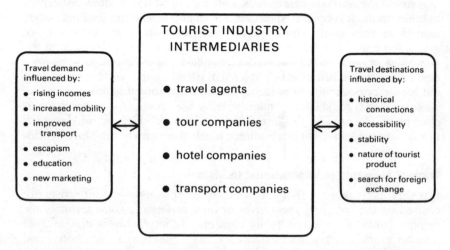

Figure 1.2   Primary elements of the international tourism industry and influences on the market

Several preliminary conclusions can now be made about the nature of international tourism and its role within the development of the Third World.

1   International tourism is unbalanced with most power and influence being held by intermediary companies controlling the metropolitan origins of Third World tourists.
2   The international tourism experience is often inequitable with foreign demands for a luxury being met by local requirements for hard currency, in circumstances where few alternatives exist.
3   Few of the factors influencing tourism in poor host countries relate to the tourist industry alone; most of them are symptomatic of a general condition of underdevelopment.
4   Few opportunities exist for Third World host countries to cut out the intermediaries and deal with their sources of tourist supply directly.

Tourism's costs and benefits

Because few of the poorest developing countries have obvious income-earning alternatives to tourism, it is sometimes adopted only as a last resort. However, with formerly valuable exports, like cane sugar, suffering from the general fall in world commodity prices and competition from larger and more efficient agricultural producers, tourism in these circumstances is a mixed blessing and demands a careful accounting of the complex pattern of costs and benefits involved. Some of the more obvious of these are found in the United Nations diagram contained in figure 1.3.

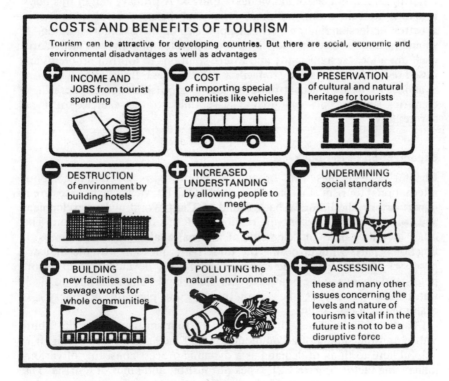

Figure 1.3 Costs and benefits of tourism
Source: United Nations Environment Program (1979), in Senior (1982)

This complex matrix of advantages and disadvantages ensures that governments must face an unenviable task of trying to weigh gains from new income and employment against certain less direct and long-term losses. It is difficult indeed for politicians to reject a new hotel project, for example, on environmental or social grounds when construction means present jobs and political prestige. In fact it is difficult to oppose any substantial foreign investment in situations where the few competing development prospects may actually have worse impacts on the host community than those associated with tourism. How can one reasonably hope to choose between the competing merits of plantation agriculture, export processing zones for manufacturing industry, or tourism, when any or all of these things are actively sought by Third World countries in a highly competitive environment?

Clearly then, there is a difficult range of questions to answer and very little likelihood of finding general solutions which will cover the widely differing circumstances of individual countries. A primary aim of this book is to introduce the extent and characteristics of the issues involved to enable a better understanding and critical appraisal of the existing situation. It should be emphasized that decisions about development in the Third World itself are a local responsibility and where solutions to tourism problems are raised in this book, it is done to facilitate this process. International tourism is almost by definition controlled by interests outside the peripheral host countries and is only marginally susceptible to the exercise of local sovereignty.

Organization of the book

Chapter 2 examines two broadly different ways of thinking about international tourism as a complex activity. One is a view which places great importance on the historical background of Third World under-development and considers today's economic realities in terms of a global system. The other identifies the main elements of international tourism as a series of functional parts without posing the underlying question why the industry possesses its present characteristics. The recent development of resort hotels and casinos in the South African 'homelands' and surrounding black states (also known as the southern African 'pleasure periphery') is examined as a case study.

Chapter 3 looks closely at the nature of tourism demand in the Third World and the supply of accommodation and facilities for the visitor. It is helpful here to think of demand as a dynamic process consisting of various influencing factors (figure 1.2), types of tourism, and kinds of tourists. The supply of tourist facilities at the destination can be thought of as an evolving process, as well as a more static activity producing built accommodation, attractions, transport, and all kinds of support services. The dramatic impact of the 1987 military coups on the Fijian tourist industry forms the case study.

Chapter 4 deals exclusively with the economic impacts of tourism and is subdivided into sections looking at some of the economic characteristics of the industry, its economic benefits, and costs. The misuse by consultants of a well-known analytical technique in a study of Caribbean tourism forms the case study. Chapter 5 is divided into two parts dealing with environmental and social impacts of the industry at Third World destinations and case studies examine visitor management in an East African game park and the effects of sex tourism in South-east Asia.

Lastly, it is important to consider the sort of national planning issues facing a host country government determined to promote tourism as a development strategy. Among the most important of these are the role of government itself and the possibilities for public or community participation in making significant decisions. The latter concern goes much further than seeking more local (Third World) participation in the decisions affecting the industry and covers ways in which ordinary people may be included in the decision-making process, because it is they who are most affected by tourism development.

Key ideas

1 Tourism accounts for some 6 per cent of world trade.
2 Most of the industry is located in Europe and North America, with only one-eighth of the market shared between the other world regions.
3 The most rapid growth in the industry in recent years has been in the Third World.
4 Tourism is an invisible export with the unique characteristic that the purchasers of its 'products' have to travel to a foreign destination in person to consume them.
5 The tourist market is controlled by 'intermediary' companies, some of them transnational in character, which have expanded to include interests in all major sectors of the industry.
6 Tourism results in a complex series of economic, environmental, and social impacts in host societies. Assessing these costs and benefits in the Third World is complicated by difficulties in measurement and a lack of local control over the industry.

2
Conceptualizing tourism's place in development

There is a need to look for general explanations or 'rules' about the way the international tourism industry works in practice if we are to understand its role in Third World development. The same applies to the other important activities making up world trade and it is essential that tourism is not studied in isolation of them. This assertion seems pretty obvious today, but it is only in the past decade that its acceptance has transformed our theorizing about tourism.

Two major approaches represent differing schools of thought in the modern literature. One view, labelled here as the 'political economy approach', is based on the premise that tourism has evolved in a way which closely matches historical patterns of colonialism and economic dependency. The shape of the industry according to this school of thought is so firmly governed by the political and economic determinants of world trade that little attention is paid to some of its other interesting features. Among the latter are the diversity of possible destinations, different kinds of holidays, or even how the tourist feels about the travel experience. The overall tone of political economy analyses tend to be negative about the effects of tourism, seeing it as yet another means by which wealthy metropolitan nations develop at the expense of those less fortunate.

The other view is much more concerned with classifying tourism in terms of its many functional parts without any political overtones. This perspective pays little attention to the historical experience of change in Third World societies and the possible contribution of the industry to present inequalities. In other words, it is a mode of analysis which attempts to put forward a neutral viewpoint in a situation where it is very difficult not to take sides. The emphases here are on the considerable economic importance of the industry to all participants and upon ways to improve its efficiency and minimize its adverse effects. This functional approach is generally optimistic, seeing most problems as capable of resolution through good management and appropriate policy measures.

It would be wrong, however, to see these two ways of theorizing in complete opposition to one another. They are both useful in helping us to fully appreciate the diversity of the subject and also represent the differing interests of the major participants in Third World tourism. We might conclude, at the risk of some oversimplification, that the functional approach is tailored to the economic objectives of the metropolitan core nations, whereas the political economy analysis sees the industry from its other end in the Third World periphery.

Political economy approach

The notion that tourism perpetuates many existing inequalities, despite its considerable economic benefit to poor countries, is closely associated with the work of Stephen Britton in the South Pacific. The political economy approach, as we have noted, probes beneath the surface characteristics of the industry in its search for the causes of problems. To do this successfully requires us to briefly examine two interrelated subjects: the way the industry is organized and the distinctive structure of Third World economies. It should then be possible to go one step further and introduce a conceptual model generalizing about the operations of international tourism.

The geographic focus of this discussion is on the so-called capitalist or free market economics and does not include socialist states where business enterprises are generally government owned. International tourism is poorly developed in Third World countries with socialist forms of government (Cuba and North Korea are representative), though this is no longer true of mainland China or even Vietnam where great efforts are being made to attract foreign visitors. Similarly, our comments on the industry in the metropolitan countries do not cover the socialist states of Russia and Eastern Europe, because there are only minor links as yet between them and Third World tourist destinations. Two examples of this small and little publicized industry are found in the growing presence of Russian holidaymakers in Vietnam and Cuba.

The organization of international tourism

Most tourism emanates from the desire of affluent middle classes in metropolitan countries to travel abroad and the companies which have emerged to service the market have organized themselves in a manner to best exploit this demand. The three main branches of the industry – hotels, airlines, and tour companies – have become increasingly transnational in their operations during the 1970s and 1980s, to the point where these large enterprises dominate all others.

Transnational hotels, for example, are characterized by certain features which are responsible for their successful invasion of Third World destinations.

1 They seldom invest large amounts of their own capital in the Third World, but seek such funds from private and government sources locally, thus minimizing risk.
2 Associated infrastructure like new roads and power supplies are essential in resort development and are similarly funded through local sources or via foreign loans.
3 A viable visitor flow is ensured through worldwide marketing campaigns.
4 Transnational corporations participate in the profits of their Third World hotels through charging management fees, limited direct investment, and various licensing, franchise, and service agreements. In all such cases the ability of the parent company to withdraw from these arrangements puts it in a controlling position.

Third World countries wishing to attract international hotels have few alternatives to this arrangement and are thus immediately locked into an unequal trading relationship.

In the case of air travel the destination countries are clearly concerned to ensure good accessibility, a share of revenue from fares and freight, and participation in important decisions about the direction, volume, and timetabling of air movements. In all these areas inequality between the transnational companies and destination interests persist. Although there are some viable national carriers in the more economically advanced peripheral countries (Thai and Indian Airlines to name but two) and very successful ones in those countries which have developed to a stage of advanced industrialization (Singapore and South Korea are representative), the majority operate only in association with 'parent' metropolitan airlines or on regional routes. In several cases attempts to establish regional airlines among groups of small countries have met with failure (as in the Caribbean and East Africa), because of political differences among participants and the sheer difficulty in keeping abreast of costs and maintenance from a peripheral location.

Transnational tour operators have revolutionized international tourism through the successful marketing of tourist packages consisting of many different items. This has raised the potential volume of sales far above that which could be expected from supplying a single service, such as a ticket or hotel room, on its own. This, in turn, has given the tour companies an increased ability to bargain with other suppliers in the industry and is leading to the sort of integration among the tourism inter-mediaries noted in chapter 1. The fact that foreign travel demand is so sensitive to price adjustments (termed a high price elasticity by economists) means that the ability to lower costs by selling seats on cheap charter flights and in reserved 'blocks' of hotel rooms is crucial.

The net effect of this organizational arrangement is noted by Stephen Britton as having three important consequences in the Third World. First, the bulk of tourist expenditure is retained by the transnational companies. In cases where the tour includes a foreign carrier but uses other local facilities only 40 to 50 per cent of the tour retail price remains in the host country. If both airline and hotels are foreign owned this drops to 22 to 25 per cent. Second, tourists visiting Third World countries tend to be increasingly confined to isolated enclaves separated from most of the local population. This is a particularly noticeable pattern in beach and island resorts and will be examined in some detail later in the book. Third, standardization of the tourist package increases the substitutability of one 'surf, sand, sun, and sex' destination by another, decreasing the ability of host countries to gain adequate control over their own visitor industry.

The structure of Third World economies

The political economy approach attempts to show how international tourism flourishes in a world economic system characterized by severe distortions and imbalances. The latter are seen as a direct consequence of imperial domination of the Third World during the past and the peculiar pattern of trading links and 'spheres of influence' established at that time. A school of thought known as 'dependency theory' originated among Third World scholars in the 1960s and 1970s which sought to explain this unequal relationship according to these historical determinants.

There is insufficient space to cover such global theorizing about economic development in any detail here except to note that metropolitan companies, institutions, and governments in the post-colonial period have maintained special trading relationships with certain élite counterparts in Third World countries. These representatives of the ruling classes gain most benefit from the less-than-equal share of income and profits which remain inside a peripheral economy. The picture is bleak for small local firms and the majority of the population who do not have good connections with the favoured few.

Thus we may portray the industry from Britton's political economy perspective as a three-tiered hierarchy with its apex in the metropolitan parent companies, connected to an intermediate level of associate firms in the Third World and having at its base a collection of small-scale local firms. The relative importance of the three levels can be seen clearly in the diagram of tourist expenditure (figure 2.1), where local firms (the small triangles) are shown as receiving only a minor proportion of overall financial benefits.

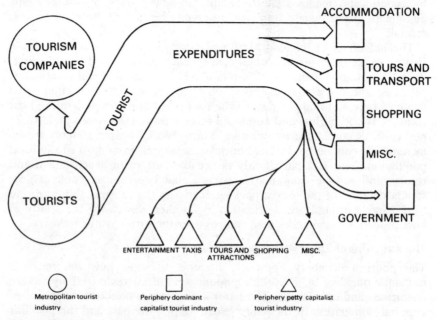

Figure 2.1   Generalized distribution of tourist industry expenditure
Source: Britton, S. (1981)

The enclave model

Dependency relations also have important physical, commercial, and social dimensions which are illustrated by an enclave model of tourist development. Britton has suggested a diagrammatic form of the model which is generally applicable to most of the Third World (figure 2.2). Here are shown the primary return flows of tourists from metropolitan countries to key cities and resort enclaves in the periphery. Some of the destinations are in Third World capitals like Bangkok, Manila, or Cairo and others are located in rural enclaves with a supporting network of local attractions. These 'environmental bubbles' are found in their most extreme form in the self-contained tourist villages pioneered by the French 'Club Méditerranée'. The promotional jingle for this company claims that 'the best things in life

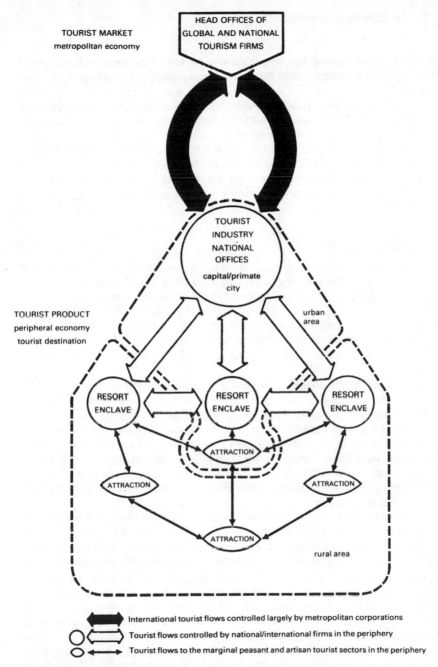

Figure 2.2 Enclave model of tourism in a peripheral economy
Source: Britton (1981)

are free', emphasizing the comprehensive nature of the tour package where even spending money inside the resorts is replaced by a system of prepaid tokens.

In summary, the political economy view dwells on the structural inequalities in world trade and suggests that international tourism is unlikely to achieve a better balance among its rich and poor participants until a corresponding shift also occurs in the whole pattern of country-to-country relationships. As we will see later, the proponents of this view are anxious to minimize the worst examples of exploitation by seeking public ownership of more of the tourist industry in host countries and direct marketing of the product where this is possible.

Functional approach

A rather different and more analytical way of generalizing about international tourism is to subdivide the travel process into three main elements: a *dynamic* phase covering movement to and from the destination; a *static* phase involving the stay itself; and a *consequential* element describing the chief economic, physical, and social impacts on the environment. These categories are illustrated by Alister Mathieson and Geoffrey Wall (figure 2.3) as a set of interconnected parts with feedback links throughout the system.

Any attempt to model a complex activity like tourism will be only partially satisfactory in representing reality and is usually done with some specific purpose in mind. The more universal the objectives, the more abstract the model is likely to be. The functional classification shown here had its origins in a study of tourism impacts and is quite successful in demonstrating how these varied consequences interrelate with other features of the process. It stops short, however, of dealing with causal relationships and an apolitical framework reduces its value in analysing the Third World situation.

A good example of the limitations which can result from using only one approach arises in the differing emphasis placed by the two models on destination (static) issues. The concern of political economists is found in the portrayal of international tourism as a means of exploiting Third World societies and is described in the operations of transnational companies, distorted expenditure flows, and the evolving pattern of resort enclaves. The functional view by contrast pays little attention to inequalities in the industry, preferring to concentrate on describing the characteristics of the tourist, various impacts and different kinds of destination. The static approach adopted in figure 2.3 is itself confusing because it neglects the dynamics of change at tourist destinations. Not only does the shape of the tourist industry in the Third World change over time, but also it can be

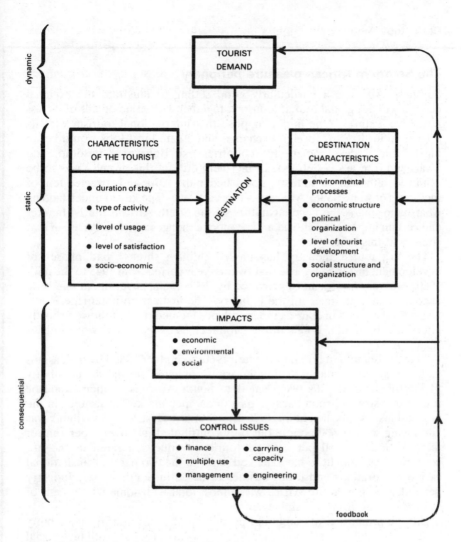

Figure 2.3   Functional framework of tourism process
Source: adapted from Mathieson and Wall (1982)

adversely affected when increasing popularity threatens things like the carrying capacity of local attractions. As we have already indicated, the full picture emerges only when both perspectives are brought together.

Case study A

The Southern African pleasure periphery

Southern Africa is a particularly good setting to illustrate the political economy of a regional tourist industry, together with some details of its less attractive features. One might suppose that international tourism with its opportunities to earn foreign exchange and create local jobs would be an ideal form of development for countries like Botswana, Lesotho, and Swaziland (the BLS countries). In fact very little of this promise was to be achieved, although the industry flourished in the 1970s and attracted tens of thousands of white South African visitors each year. The key to understanding what happened is the peculiar nature of the South African tourist market and certain important political and economic changes which occurred in that country at this time.

The BLS tourist industry has passed through three broad phases of development historically. The first twenty-year period from 1945 to the mid-1960s saw very few facilities provided by the colonial government and little encouragement to invest in the industry. The first casino-hotel (the Royal Swazi Spa) opened in 1965 as independence loomed, to be followed shortly afterwards by similar hotels in the capital cities of the other two countries (figure A.1).

There was a tourist boom in the late 1960s to mid-1970s led by the casinos and their special brand of 'forbidden fruits', of gambling and the possibility of inter-racial sex. This was denied to South Africans in their domestic tourist industry and soon became part of an open marketing strategy in the BLS casinos. At its height in 1976 the tourist industry in Swaziland was employing some 1,600 people but only contributed just over 2 per cent of GNP. As much as 60 per cent of tourist revenue generated in the BLS countries was thought to have 'leaked' back to the two parent transnational casino corporations and other businesses based in South Africa. Industry dependence upon South Africa was almost total with some 80 per cent of visitors coming from this single source.

But the promotion of gambling, pornographic films, and the availability of sex across the colour bar was leading to a form of tourism with high social costs. As many as three-quarters of the slot-machine clientele, for example, were local people whose very low incomes could support such habits only at the expense of basic needs like food and clothing. An assessment of tourism in the BLS countries which neglects such social costs is guilty of a biased and very limited appraisal of the industry's true role in national development.

A steep decline in the fortunes of BLS tourism began in the late 1970s with

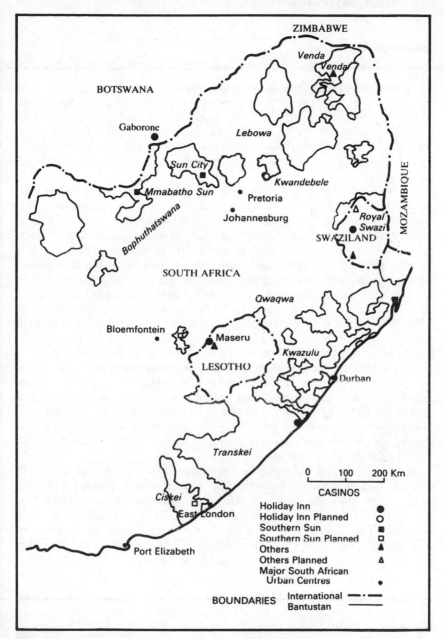

Figure A1   Casino development in Southern Africa 1983

Source: Wellings, P. and Crush, J., 'Tourism and dependency in Southern Africa: the prospects and planning of tourism in Lesotho', *Applied Geography* (1983) 3:219 (reproduced by permission)

Case study A (*continued*)

occupancy rates falling by a quarter in only two years. The reason for this had little to do with events inside countries like Lesotho and Swaziland where the physical capacity of the industry was still growing. It had everything to do with the South African source of the visitor flow where a big increase in petrol prices had dampened demand, and a major change in marketing strategy by the tourism transnationals had altered the shape of the industry.

By 1981 South Africa had granted its own form of 'independence' to four out of its ten black homeland states, but none of them had secured any recognition internationally. Just as a decade earlier tourism had beckoned as an economic saviour in the BLS countries, it now seemed to offer poor South African homelands a welcome new source of income. With Pretoria's encouragement, the transnationals built new casino-hotels in the homelands, effectively diverting many tourists from the BLS countries to attractions much closer to home. The device which made this possible was the ability to offer 'forbidden fruit' attractions in homeland locations which were now self-governing and only partially subject to South African law. The fact that new casino-hotels were also a prime means of bringing South African private investment into the homelands was a bonus for the government.

The South African casino industry restructured itself in 1983 to maximize the benefits arising out of the homelands developments. A single giant corporation called Sun International Ltd took over all the casinos except for the Maseru Hilton in Lesotho. The effects of single ownership will minimize competition and enable the owners to concentrate on the most profitable sections of the market. This, in turn, presents a difficult development dilemma for the BLS countries.

On the one hand a reduction in the number of homeland casinos could redirect business to the BLS countries leading to a revival of their hotels but, on the other hand, this may serve only to rekindle an industry which has done very little for the economic independence of these small countries in the past. The debate thus returns to the overall role of tourism in development and raises the possibility of alternatives, such as the encouragement of industries with lower social costs.

Key ideas

1 Ways of conceptualizing tourism vary according to factors such as the specific purpose of the analysis, the background, motives, and beliefs of

the researchers, and the level of explanation or description sought.

2   Basic differences in approach exist between a political economy model of international tourism, which seeks to explain present conditions in terms of evolving global relationships between rich and poor countries; and the functional view which aims to describe and classify separate elements of the tourism process.

3   Political economy approaches are generally negative about tourism's overall contribution to Third World development, considering economic benefits to be outweighed by its other consequences.

4   Functional views do not consider the tourist industry in parallel with other branches of world trade, nor with an historical perspective. Implicit in this approach is that tourism is generally desirable and that problems can be resolved by adopting appropriate practices.

5   Distinctions between the two approaches echo similar differences found in the analyses of other aspects of Third World development. Among these are migration, housing, and employment studies.

3
Tourism demand and the supply of facilities

Tourism demand is a broad term which we will use to cover the discussion of four more specialized topics: the factors governing levels of demand; its spatial characteristics; attempts to create typologies of different kinds of tourist activities; and tourist decision-making. It is also an imprecise term consisting of three main components, actual demand from those who become tourists; potential demand of those who would like to travel but cannot for reasons of time or money; and deferred demand of those who could travel but do not for some good reason. Potential and deferred categories are difficult to measure and rarely taken into account in practice, even though knowledge about them would be important in forecasting the size of future tourist developments.

Supplying tourist facilities at Third World destinations is a dynamic process of growth and change, as well as the static provision of a range of buildings, services, and attractions. We will consider some models of the way in which tourism develops over time separately from the static elements of supply like accommodation and transport. A case study of the dramatic effects of the 1987 military coups on Fijian tourism illustrates how the prospects of a successful industry can change overnight and how damage to the economy will also serve to further weaken the position of the very poor.

Factors governing levels of demand

There is considerable confusion about the treatment of tourism demand because some researchers make no distinction between economic, social, and technological influences of the kind we noted in figure 1.3, and other important factors like the effects of tourism promotion or even the personal motivations of individual travellers. The former factors will be considered here and the question of travel motivation left until later in the chapter. A further point to note is that some special influences on demand originate in the Third World destinations or are connected with specific aspects of travel. There is increasing recognition, for example, that special circumstances are involved where terrorism, political instability, or uncertainty intervene to threaten health and safety. Recent instances of these effects range from a virtual boycott by travellers of certain airports and airlines, to the abandonment of formerly popular tourist destinations. Perhaps the most notorious example in recent times comes from the central African country of Uganda, where tourism was the third most important source of foreign exchange before the beginning of the Amin regime in 1972. Just six years later visitor numbers had dropped from 85,000 to only 6,000 and most of the country's hotels were destroyed or vandalized.

The major economic, social, and technological determinants of tourism demand, however, are firmly located in the metropolitan countries. High and rising incomes, increased leisure time, good education, and new and cheaper forms of transport are all found in these economies, and it is no surprise to learn that they are also the suppliers of most of the tourists visiting the Third World. Not quite so obvious is the extent to which sophisticated promotion of the tourist product has created a demand which did not previously exist. In part this involves the marketing of packaged tours but tourism promotion also means creating the image of a destination in the mind of the potential traveller.

Tourism imagery can be looked at in two ways, it is both a personal process, which helps to determine what sort of holiday or trip to take, and also a deliberate part of marketing strategy by tour companies. This has led to the growth of a mythology about some Third World destinations which seems designed to attract visitors by creating an unreal picture of foreign places. There are limits, of course, to complete deception because this would soon result in no visitor traffic at all. Imagery is thus an essential part of tourist decision-making but it can be manipulated to suggest that a destination has all the requirements of the 'bliss formula'.

An extraordinary African example (table 3.1) shows separate tourist advertisements for holidays in two different countries in 1976. They use identical language to describe quite different cultures in totally different regions of the continent. The sales image for an African holiday is standardized here with the names of different countries and certain special items being included or removed from the advertisement at will. Thus one strange tropical place may be substituted by another according to marketing strategy.

Table 3.1 Tourism imagery and African holiday destinations

Thomas Cook's advertisement	*Kuoni Travel booklet*
'Visit Kenya's sun-drenched coast with idyllic beaches lapped by the clear enticing waters of the Indian Ocean . . . where the pulsating rhythm of tribal drums and dances can still mystify and thrill; unbelievably, besides this scene from a Tarzan epic the glorious golden beaches shimmer in the sun and hotels afford the comforts from which to sunsoak and explore.'[1]	'The Gambia is an exciting and intriguing miniature of the dense jungles and great rivers of Africa. . . . A place were the pulsating rhythm of tribal drums and dances can still mystify and thrill; where hippos and crocodiles are still at home. . . . Unbelievably, besides this scene from a Tarzan epic, the glorious golden beaches shimmer in the sun and hotels afford simple modern comfort from which to sunsoak and explore.'[2]

Sources: adapted from (1) Geshekter, C. (1978) 'International tourism and African underdevelopment: some reflections on Kenya', in M. Zamora, V. Sutlive, and N. Altshuler (eds) *Tourism and Economic Change*, Williamsburg, College of William and Mary, 57–88; and (2) Harrell-Bond (1978)

Tourism demand and its particular effects on the attractiveness of any single Third World destination are made up of a large array of plus and minus factors (table 3.2). The most important of these concern the cost of the trip, which is itself a combination of various things; but of particular significance are the exchange and inflation rates in the host country. During the early 1980s, for example, a low rate for the US dollar coupled with high inflation resulted in raising domestic prices in Mexico some 20 per cent above those in the USA, with a consequent falling off in visitor traffic.

Other factors shown in the figure are much more difficult to quantify and are unpredictable. Natural disasters, for example, are a recurrent hazard in hurricane or cyclone-prone island resorts in the tropics and can set back or cripple a successful tourist industry overnight. Some of the better known places which have suffered in this way are Jamaica, Fiji, Vanuatu, and the Cook Islands. Bad publicity arising from the treatment of Third World tourist destinations in the foreign press is found where visiting journalists have little familiarity with local conditions. Reporting of the communal violence in the French territory of New Caledonia in late 1984, for example, is an apt illustration of this effect and 'ensured a superficial, simplistic and unbalanced coverage' (Connell 1987).

Spatial characteristics of demand

It was indicated in chapter 1 that three of the eight world tourism regions (West and East Europe and North/Central America) account for seven-

Table 3.2   Assessing the characteristics of a tourist destination

Advantages	Disadvantages
physically attractive	high inflation
good climate	strong currency
geographically proximate	high crime rate
low-cost travel	incidence of terrorism
good facilities	incidence of natural disasters
politically stable	politically unstable
economically prosperous	unpopular government or regime
cultural, social, historical ties	bad publicity
new, exciting location	economically weak
cheap accommodation	well-tried location

Source: Senior (1982)

eighths of all arrivals (figure 1.1). The remaining five regions (South America, North Africa and Middle-East, Africa, Far East, and Oceania) cover most of the Third World and attract only the small balance of about one-eighth of the total market.

The pattern of movement between and within these travel regions has been examined for 1979 by Robert Senior with some interesting results. In general, most international travellers stay within their regions of origin but the following details give a better idea of the situation in those parts of the world with a significant number of tourist destinations.

1   Central/North America: by far the largest number of visitors to the Caribbean come from Canada and the USA, with some island countries receiving over 90 per cent from this source (that is Bermuda and British Virgin Islands). The single largest destination is Mexico which receives more than 80 per cent of arrivals from the USA.
2   South America: the majority of foreign visitors come from other countries in the region (over 75 per cent in the cases of Chile, Colombia, and Paraguay). Most overseas arrivals go to Brazil, Venezuela, Bolivia, Peru, and Ecuador.
3   North Africa and Middle-East: most foreign visitors are from Europe and there is only a small proportion of local tourists.
4   Africa: has the widest spread of arrivals but great variations exist for individual countries, ranging from 90 per cent local tourists in Botswana (mainly from the Republic of South Africa) to only 3 per cent local in the Ivory Coast. Overseas sources of tourists vary strongly on a regional basis within the continent, with the French forming up to 40 per cent in some West African countries; 80 per cent visiting the Gambia come from Scandinavia, and large numbers of British and West Germans go to East Africa.

5   Far East: regional visitors are in the majority in Macau, Malaysia, and
    Taiwan but are less significant in Singapore, Hong Kong, India, and
    Japan. Major sources of overseas tourists are Australia, Japan, Britain,
    and the USA.
6   Oceania: the chief sources of local travellers are Australia and New
    Zealand but the region is rapidly becoming dominated by overseas
    visitors from the USA and Japan. The most popular destinations
    (Hawaii, Australia, and New Zealand) are not part of the Third World.

Tourist typologies

It is also possible to think of patterns of tourism demand in a more
qualitative way which focuses on different kinds of tourists and various
forms of travel. In early books and articles the traveller was often thought of
as a rather 'superficial nitwit', easy to please and lost in an environmental
bubble of hotels and entertainments. This view was challenged in the early
1970s when the tourist acquired a much more respectable status as the
'modern pilgrim' in search of the symbols of human culture. Tourism was
even likened to a new sort of religion.

Whatever our overall interpretation of tourism's role in society, it is clear
that the term is now used as a label to cover many different kinds of
travellers and some researchers have suggested typologies to assist us in
describing them. The basis upon which they are created and the purposes
intended have given rise to distinctly different typologies. There is a general
distinction between an 'interactive' kind stressing the great variety of
travellers and their behaviour at a destination, and so-called 'cognitive-
normal' models focusing on the causes of travel. An example of the former
was devised by Valene Smith and separates tourists into seven demand
categories:

1   Explorer: very limited numbers looking for discovery and involvement
    with local people.
2   Elite: special individually tailored visits to exotic places.
3   Off-beat: the desire to get away from the crowds.
4   Unusual: the visit with peculiar objectives such as physical danger or
    isolation.
5   Incipient mass: a steady flow travelling alone or in small organized
    groups using some shared services.
6   Mass: the general packaged tour market leading to tourist enclaves
    overseas.
7   Charter: mass travel to relaxation destinations which incorporate as
    many standardized western facilities as possible.

Each of these categories has a corresponding range of impacts on the host society and destination, with progressively more intensive effects being felt in each category down the list. Thus tourists on a hiking holiday in Nepal will have minor social and physical impacts on their hosts compared to the effects on the Balinese of the huge numbers of young westerners who visit this beautiful Indonesian island on packaged trips.

A cognitive-normal typology suggested by Erik Cohen seeks to distinguish visits in terms of what they mean for the traveller. Here the typology covers visits directed at pleasure alone as well as those where there is an emphasis on pilgrimage to some new and personal experience:

1 Recreational: one of the commonest forms of tourism where the trip is designed to relieve the strains and tensions of work with no deeper significance involved.
2 Diversionary: when the visit is a pure escape from the boredom and routine of home life.
3 Experiential: describes the tourist as the modern pilgrim looking for authenticity in the lives of other societies because he has seemingly lost his own.
4 Experimental: when the traveller begins to experiment with life-styles other than his own.
5 Existential: describes the tourist who actually acquires a new spiritual centre as a result of the travel experience.

Such generalizations help us to view tourism both from personal (what it does for me) and host society (how do we judge what to accept) perspectives. There is considerable evidence, for example, to show that the personal aspirations of western tourists may not accord with the priorities held by Third World governments for the development of the industry.

Some governments wish to maximize income from the industry by encouraging mass tourism with a minimum of local contact (as in new beach resorts in Mexico); others wish to change their tourist trade up-market to gain similar returns from a smaller number of top-spending visitors (as reported recently for Bali); while still others appear to desire mass tourism with maximum visitor–host interaction by encouraging the use of village accommodation and hotels. The Minister of Tourism in the South Pacific country of Western Samoa, for example, has recently been quoted in the Australian press as saying, 'In the past we've been very reluctant to accept tourists. It was the fear that money would corrupt culture. . . . I want 3000 hotel rooms in Apia [the capital] for tourists coming in on three or four 747s per day.' The idea is that Apia's hotels would serve a transit role for tourists fanning out into the country's 360 villages as paying guests.

There is clearly an endless variety of possible combinations, but we should not forget the severe difficulties facing any Third World society wishing to change market trends to meet new national goals, when such trends are determined by powerful external forces. As noted in chapter 2, there is always another poor and warm destination country looking for an entry into the competitive world of international tourism.

Decision to travel

It is evident that plus and minus factors affect demands by tourists to visit overseas destinations and that the traveller ultimately creates an image of a desirable place to visit. The question of how that decision is made introduces us to the psychology of tourism.

The behavioural characteristics of tourists (motivations, attitudes, needs, and values) all combine in a complex way to influence travel decisions and one of them, motivations, is selected here to illustrate the diversity of issues involved. Four main motivational categories are identified in the literature and summarize a very long list of separate issues. First there are physical motivations which cover the obvious health, refreshment, sport, and pleasure desires of many travellers; second are cultural motivations, that is curiosity about foreign people and places and a desire to participate in international events like the famous carnival in Rio de Janiero or the Oktoberfest in Munich; third are the personal motivations such as visiting friends, relatives, a desire for change, a spiritual experience, and the sheer romance of travel to distant places; and finally there are prestige and status motivations which encompass hobbies like flying or sailing, education, conferences, and 'keeping up with the Joneses'. Together these influences help to create the destination image which in much of the Third World has become stereotyped as a search for the four S's of surf, sand, sun, and sex.

The actual process of decision-making is thought to be a behavioural process whereby the traveller acts on the basis of limited knowledge to find the most satisfactory outcome to meet his needs. This is rather different from a familiar model used in economic studies which assumes that the decision-maker has perfect knowledge and acts in a rational way. The average tourist is faced with considerable uncertainty and may only have a scanty appreciation of a distant destination. Five sequential phases or steps are identified by Mathieson and Wall as being involved in coming to the travel decision.

1 Travel desire: the initial period when a need to travel is felt and when the pros and cons are weighed up.
2 Information collection and evaluation: involves the process of finding out about the trip from travel agents, books, and acquaintances. Information is evaluated against cost and time constraints, alternative possibilities, and many other factors.

3   Travel decisions: covers the destination, way of travelling, accommodation, and activities involved.
4   Travel preparations and experience: involves tickets, bookings, travel money and documents, clothing, and travel itself.
5   Travel satisfaction evaluation: the whole experience is constantly evaluated before, during, and after completion and the results used to influence future decision.

There are limited opportunities for destination countries to influence personal decision-making in a direct way, though an ability to do this could greatly increase tourist traffic. One of the best recent examples is the appearance of the popular Australian comedian Paul Hogan in a successful series of US television advertisements promoting tourism in his country. Third World government tourist offices, airlines, and some resorts employ similar techniques to influence the crucial first phase in decision-making but are restricted by the high costs involved and the difficulty in clearly differentiating their tourist product from that being offered by competitors. Two very different ways of travelling to the remote and beautiful French Polynesian island of Moorea are shown in plates 3.1 and 3.2.

Plate 3.1   Yachts in Captain Cook Bay, Moorea, French Polynesia

Plate 3.2   Tahiti airport with island of Moorea in background

Supply of tourist facilities

Theorizing is not confined to modelling tourism's role in global development
and making typologies of different kinds of tourists, but can also further our
understanding of how the industry evolves at a Third World destination.
The enclave pattern identified in chapter 2, for example, did not appear at
the same time in all host countries nor does it characterize all resorts. Some
researchers have sought to explain these anomalies by examining change
over time as the determining factor, while others have based their
generalizations on more complex ideas of the nature of change itself.

A linear model

A simple time sequence has been suggested, consisting of three stages of
tourist industry development in a destination society.

1 Discovery: when the new place begins to attract attention in the
   metropolitan source countries.
2 Local response and initiative: when national entrepreneurs respond to
   new income-earning opportunities from tourism.
3 Institutionalization: when the industry is taken over by the large foreign
   companies.

This idea, based on experience in Indonesia, suggests a common pattern exists in many countries and that the relationship between tourist and the host community also changes over time as the industry evolves. Not all countries follow this pattern, however, and it is accepted that some may experience stage 2 or stage 3 development from the outset. A good example of this comes from the Republic of the Maldives where the expansion of luxury resorts on more than 40 small Indian Ocean islands followed the construction of a new airport capable of handling wide-bodied jets. This could be called a form of institutional development with relatively minor participation from local Maldivian sources.

A multi-linear model

This idea, attributed to Erik Cohen, recognizes that Third World tourist destinations have probably developed in several different ways, with one determining factor being the way in which tourism was introduced. On the one hand, when tourism grows organically within the system the development of the industry from small beginnings is likely to follow the linear sequence noted above (as in India or Thailand); on the other hand, when tourism is induced from outside the first stage of contact is the most institutionalized, when a wide gap between visitor and hosts still exists (as in the Maldives or the Seychelle islands in the Indian Ocean). It seems that where tourism is externally induced from the beginning, as in Fiji, a reverse pattern develops with groups in the host country attempting to take over and de-institutionalize the industry as they become more familiar with it. But even if this does occur, little real control is likely to shift from the powerful tourism intermediaries to local people. Another possibility is the emergence of Third World-based transnational hotel chains like the Indian Oberoi group which have the ability to penetrate the metropolitan countries from the periphery.

The elements of supply

There are five broad types of facilities and services noted in the literature which are found in tourist destinations worldwide:

1   Attractions: these may be classified into natural (land forms, flora and fauna) and man-made (historic or modern) or by cultural distinctions such as language, music, and folklore.
2   Transport: there is a close relationship between tourism growth and transport developments. Some Third World destinations and certain locations within these countries, for example, are favoured by easy access to world air routes.
3   Accommodation: this is subdivided into a commercial sector (hotels, guest houses, holiday camps), the private sector (private residences and second homes), and camping/caravanning. There is less of a trend in the Third World towards the provision of self-contained accommodation for

tourists than in the developed countries because of relatively low labour costs.
4   Other facilities and services: this covers a large array of supporting services such as shops, restaurants, banks, and medical centres. Provision ranges from the self-contained resorts like Club Méditerranée to a total reliance on local facilities. The latter is usually found only in urban locations in the more developed Third World countries.
5   Infrastructure: this is a broad term used to cover support items necessary to provide the facilities listed above (roads, rail, airports, electricity, sewage disposal, and so on). They are usually provided by governments because of high capital costs and often serve the local population as well.

Case study B

Tourism and the 1987 Fiji coups

Tourism earned £176m. for Fiji in 1986, the largest single source of foreign exchange, followed by sugar (£127m.) and gold (£37m.). Its importance to the national economy can scarcely be overestimated yet the viability of the whole industry seemed to be threatened by the military coups which took place in May and September 1987. Early reports in the Australian press were dramatic with sources in the industry being quoted as saying, 'We are incredibly concerned. We are very, very worried. . . . It is the worst thing we have had happen, particularly when you consider there has been no violence in tourist areas and even the violence in Suva [the capital] was confined to only one day, really.'

Cancellations soon had tourism leaders warning that they would be lucky to achieve half of the anticipated visitor target of 285,000 for 1987 (table B.1). Reports came in that many of the large resorts were almost empty and that local staff were being asked to take all due leave and the working week was being reduced to four days to prevent redundancies. In Australia, where some 43 per cent of Fiji's tourists originate, prices of packaged tours were slashed in the hope of filling empty beds (plate B.1). Local resorts in northern Australia at a similar latitude to Fiji reported booming business with a survey of disappointed would-be tourists showing that 83 per cent of them did not intend to postpone their holidays but would go elsewhere. Switching to Queensland would save the intending tourist 25 to 50 per cent over an equivalent holiday in Fiji, it was claimed.

However, a lack of continuing violence in the weeks following the May coup together with the enticement of huge savings on cut-price Fiji holidays soon reversed the trend. Sydney newspapers came out in June 1987 with

Case study B (*continued*)

Table B.1   How the 1987 coup hit Fiji's economy

Visitor arrivals	June 1986	June 1987	% drop
Australia	7,407	1,652	77.7
USA	3,931	833	78.8
New Zealand	1,963	510	74.0
Total arrivals	17,969	5,120	71.5

Economy	Before coup	1 August 1987
Foreign reserves	$F 195m	$F 107m
Government bonds	15%	20–30%
Fijian/Australian dollar rate	$A 1.312	$A 1.074
Interest rates	13–14%	16%
Savings deposits	$F 396m	$F 415m

Source: *Sydney Morning Herald*, 1 August 1987

---

headlines like 'Bargain-hunters rush Fiji holidays' and 'Gold Coast [Queensland] loses out in the rush of Australian tourists to post-coup Fiji on special fares'. The deals appeared irresistible, with return fares of only £210 on Air Pacific (Fiji's own airline) being far cheaper than internal travel within Australia and resulting in 900 passengers from Sydney in one weekend alone. A £1.08m. advertising campaign was launched by the Fiji Visitors Bureau in an attempt to restore consumer confidence and reverse the view of many Australians that Fiji was a place of racial tension and no longer suitable for family holidays.

In spite of this rapid reversal the fact remains that it was achieved at a considerable price. By early September 1987 the secretary of the Fiji Trade Union Council was reported as saying that 80 per cent of staff in the hospitality industry had been sacked and that public servants had taken a 15 per cent cut in salaries. The Fijian currency was devalued by over 17 per cent shortly after the May coup and this, coupled with reduced tourism receipts, is forecast to boost inflation and seriously affect an economy reliant on a high percentage of imports. It will have an adverse impact on all sectors of Fijian society and especially the 'other' Fiji, the one that tourists rarely see where grinding poverty, racial tension, and hopelessness is the norm. The following extract from a report in the *Sydney Morning Herald* of 4 June 1987 reveals one of the dark sides of 'paradise':

The well-nourished tourists have always passed by the realities, even

Case study B (*continued*)

Plate B.1   Fiji travel poster displayed in Sydney, Australia, two weeks after the May 1987 coup
Source: Visa Travel Agency, Sydney

though one jogging couple, spotted on Suva's Victoria Parade during the height of racial violence, found themselves overtaken by Indians sprinting from Fijians.

But at Jittu Estate, where Mr Dharn Raj Naiker, 50, a Hindu, kept guard over the small home he built there as a squatter 19 years ago, the other, the real, Fiji came into view [plate B.2].

Case study B (*continued*)

Plate B.2   Mr D. R. Naiker, his wife and family, Jittu Estate in Suva, Fiji, 1987
(*Photo*: Bruce Miller)

If Indians had earned more money than Fijians during more than a
century's presence in the island nation, thus earning the resentment of the
native people, there were still extremes of inequality within their ranks –
and Mr Naiker, his wife and five children are at the bottom of the heap.

The estate, with about 335 Indian families and an enclave of about 15
Fijian families, has a total of several thousand inhabitants. Nobody quite
knows how many. The Fijians and Indians have rubbed along, but there
has always been bickering.

As is the case through the Third World, there is a tendency to produce
large families, so that adults, as they get older, can look forward to
children to make up for the lack of government social welfare benefits.
But the children make them all poorer. They are caught in a vicious cycle,
even if, as one Fijian woman said, it is still more financially secure to live
in Suva than face the hopelessness of village life. . . .

Mr Naiker, a cook at Suva's Colonial War Memorial Hospital, earns

Case study B (*continued*)

> A\$56 (\$F68) a week . . . which is poor even by Suva's standards. But like
> so many of his neighbours he has made the best of it. . . . While Indian
> families on hilltops not far away were able to fortify substantial dwellings,
> on Jittu Estate the Indians were vulnerable. Three had already been
> admitted to hospital with head and body injuries after Fijians stormed the
> estate.
>
> 'Oh, we would love to get out,' Mrs Naiker said. 'But no, we've never
> been able to afford it. We were hoping our eldest boy would get a scholar-
> ship to study engineering in Australia, but he was not successful.'

Key ideas

1 The chief social, economic, and technological determinants of tourism
   demand for Third World destinations are found in the metropolitan
   countries where most travellers originate. They include factors such as
   high incomes, increased leisure, good education, and new and cheaper
   forms of transport.
2 Tourism demand is also affected positively and negatively by Third
   World destinations. Would-be travellers are attracted by low costs, good
   facilities, stability and safety but repelled by natural disasters, terrorism,
   instability, and high costs.
3 A major strategy in tourism marketing is the creation of positive tourist
   images about a Third World destination.
4 Third World destinations attract about one-eighth of the total travel
   market. Most tourists remain within their world regions and considerable
   differences are found in visitor origins among regional groupings of
   Third World countries.
5 Tourists can be broadly organized into typologies based either on the
   behaviour of visitors at a destination and the character of the visit, or on
   the motives for travelling and what is expected from it.
6 The decision to travel is a complex process based on motivations,
   attitudes, needs, and values.
7 The supply of tourist facilities and services at a Third World destination
   can be seen as a changing process over time as the industry adapts to
   market trends, as well as the static provision of certain essential
   elements (attractions, transport, accommodation, and so on).

4
Tourism's economic impacts

In 1973 the eminent Tanzanian academic Issa Shivji wrote in the Preface to his book, *Tourism and Socialist Development*, that 'The justification for tourism in terms of it being "economically good" . . . completely fails to appreciate the integrated nature of the system of underdevelopment'. It introduced a debate which illustrated clearly that tourism and economic development need to be examined at two different levels simultaneously. Ignoring either of these levels results in an unrealistic assessment of tourism's economic impacts.

First is a broadly based discussion of the political economy type which weighs economic benefits from tourism against a wide range of perceived costs. This kind of analysis rests more on arguments in political science and sociology than economics, and 'costs' include many that are hard to measure numerically. Crush and Wellings (1983) have termed it 'the seemingly endless debate on the pros and cons of tourism. . . . Proponents will rue and critics will embrace externally dictated trends over which planners . . . have very little control'. Nevertheless, the justification for this kind of discussion rests on the fact that to look only at purely economic considerations means isolating tourism from its development context in the Third World.

The other level of economic analysis is much more specific in its examination of measurable indicators of economic performance. The techniques used are sophisticated and quantitative and belie the fact that the complex interactions of tourism with other sectors of the economy are not fully understood. This approach, essential though it is for the detailed appraisal of tourism developments, is not broad enough to form the sole basis of judging the overall worth of the industry. Unfortunately, proponents of either level of debate have tended to allow their disciplinary backgrounds to influence the direction of their conclusions. Thus the political economy view is usually pessimistic about the long-term advantages of tourism investment, whilst detailed economic analysis is usually at pains to underline the benefits. It is also generally the case that studies of specific economic impacts are sponsored by government or the tourism industry itself, whereas the broader and more theoretical view usually comes from outside observers.

In this chapter we first examine the well-documented tourism debate in Tanzania before listing some of the main economic criteria found in the development studies literature. A celebrated critique of the misuse of the economic multiplier in Caribbean tourism forms the case study.

General debate on tourism and economic development

In 1970 a group of students at the University of Dar es Salaam contributed a long article about the economics of tourism to a local newspaper. It was the start of a three-month debate among academics, townspeople, and civil servants which gave a thorough airing to claims about the economic benefits of tourism, and provides a good example of the political economy perspective.

Foreign exchange earnings

The students maintained that figures usually quoted for foreign exchange earnings from tourism were gross amounts and did not take into account the real cost of imports found in the industry. The resulting net total was much smaller than usually claimed. They also questioned the importance of foreign exchange earnings as a yardstick to use in judging the value of an economic activity.

A rise in the gross national product

Many economic analyses claim that a 'multiplier' effect follows investment in tourism. Thus a tourist purchases handicrafts in a local shop and part of the proceeds of the sale becomes income to the shopowner. The latter buys food with it at another store, some of which may 'leak' out of the country if imports are involved, but the residue is income to the foodstore owner. This, in turn, is re-spent, and so on. The operation of this economic multiplier is said to raise the entire GNP and will vary from country to country and on a regional basis. We will look at an example of these claims in the case study C later in the chapter. The students questioned whether a simple rise in GNP is really a sign of development at all and whether benefits from the multiplier would be distributed fairly. They also wondered if the benefits would be of a once-only kind following new investment and not be sustained in the long term. These possibilities they called economic growth without development.

A revenue earner for government

This argument refers to the small share of tourism's economic benefits gained by a government from taxes and duties on things tourists use and buy. Thus duties on alcoholic drinks and petrol and a hotel room tax constitute an extra source of national income from the tourist industry. The figures for Tanzania suggested that for every tourist dollar, 40 cents goes on imports, 40 cents to private hotel and other businesses, and only 20 cents to the government in the form of taxes.

A generator of employment

Here the students questioned the real value to the nation of the jobs coming from tourism and compared this unfavourably with an equivalent investment in a labour intensive textile factory. According to this comparison something like twenty times the number of new jobs would be created by investment in textiles rather than tourism.

An improvement in social services

The question here is whether social services like public health, sanitation, and housing are positively stimulated by tourism. According to the students there was little evidence of it and the most likely services to be stimulated were entertainment and leisure facilities like swimming pools, casinos, and night clubs.

An American economist, Frank Mitchell, based in Kenya at this time, joined the debate in an article in which he attempted to contradict each of the criticisms listed above. Although also concerned with effects on the total economy, he appears to have made the assumption that tourism is a good thing if the economic indicators are positive. He criticized the students for their vague terminology and sought to show how economic analysis paints a much brighter picture of the industry's role in development.

On the question of a gain in the GNP, for example, Kenyan measurements were used to show a minimum net benefit to the Tanzanian economy of about 15 per cent from direct tourist spending. When it came to a textile factory rather than tourism, Mitchell pointed out that the produce from a small industry of this kind would cost Tanzanian farmers much more than they could gain from selling their cotton on the world market and buying imported cloth. Similarly the questioning by the students of tourism income distribution should also be applied to any other modern industry in Tanzania. According to Mitchell much of the national gain from tourism comes in the form of government taxes and there is nothing to stop the Tanzanian government from spending this for the benefit of the poor. Moreover, many tourist jobs are in the rural areas where modern sector employment is scarcest and could thus be said to be worth more than their equivalent in the city.

The essential difference between these points of view was summarized in the same newspaper by A. P. Mahigu, a political scientist at the University of Dar es Salaam:

> In assessing the economic implications of tourism we need to look at it from an overall view of our economic philosophy and not from the narrow and short term view of whether it makes profit or loss at a given period of time. That is where lies my disagreement with Mr Mitchell. He implies in his articles that profitability of an industry should be the sole criteria in determining investment priorities and development strategies regardless of the economic and political implications. But I maintain that investments in our country ought to be geared towards transferring our economy into a self-reliant one even if such projects may not yield immediate profits in the short-run.

Thus the economic efficacy or otherwise of tourism is transformed into a commentary on the whole condition of underdevelopment.

Mathieson and Wall have suggested that tourism makes its most valuable economic impacts in the early phases of development in a Third World country, and will diminish in significance with the coming of industrialization. They also warn against over-reliance on tourism and urge that profits from it be channelled into other sectors of the economy. Accurate though such comments are for some countries, particularly the newly industrializing ones, there is no doubt that many more will see little economic development of this kind and will have to rely on tourism, warts and all, for as long as they can.

Case study C

Use and misuse of the tourist multiplier: a Caribbean example

In 1969 a consulting firm called Zinder and Associates employed by the United States Agency for International Development published a report called *The Future of Tourism in the Eastern Caribbean*. It contained a highly controversial calculation of the tourist multiplier for these islands but gained considerable acceptance by governments in the region. Faulty use of the tourist multiplier greatly exaggerated the secondary benefits to be gained

Case study C (*continued*)

from promoting the industry and soon became the object of critical analysis by other economists. Before describing what went wrong we will briefly redefine the multiplier effect and note some qualifications regarding its use.

The tourism multiplier effect is neatly defined by Douglas Pearce (1981) as 'the way in which tourist spending filters through the economy, stimulating other sectors as it does so'. It is actually the specialized application of a general economic technique first developed by the famous British economist Keynes in the 1930s and is subdivided into three categories. These are summarized well in several examples contained in Pearce's book:

Sales and output multipliers: measure total sales or output stimulated by an initial expenditure as a ratio. Thus, $100 spent by a tourist on a meal could result in a second round of $50 spent by a waitress out of her wages on a dress, and another $25 in a third round by the dress-shop owner on weekly groceries. The total of $175 set against the $100 originally spent gives a multiplier of 1.75;

Income multipliers: are a little more complex and measure the relationship between tourist spending and subsequent changes in income in the following way:

$$K = A \times \frac{1}{1 - BC}$$

where A = percentage of tourist spending remaining in the region after some has 'leaked' away; B = percentage of income spent by residents on local goods and services; C = percentage expenditure of residents accruing as local income (after leakages).

So, if 50 per cent of tourist spending remains after first-round leakages, and 60 per cent of income is spent locally, and 40 per cent is local income, then the income multiplier is:

$$0.5 \times \frac{1}{(1 - 0.6 \times 0.4)} = 0.65$$

Employment multipliers: are the ratio of direct and secondary employment generated by additional tourism expenditure to direct employment alone. So, if 100 new jobs in the tourist industry gave rise to 20 more, the multiplier would be 120/100 or 1.2.

Case study C (*continued*)

Plate C.1   Cruise liners in Ocho Rios, Jamaica

The Caribbean example

The Zinder Report was subjected to a detailed critique in two articles in the West Indian journal *Social and Economic Studies* by Levitt and Gulati; and Bryden and Faber which form the basis of this case study example. Zinder and Associates claimed to have calculated a tourist income multiplier for the Eastern Caribbean islands of 2.3 which was derived as follows:

An estimate of $1,000 of tourist expenditure comprises:
$315 spent on accommodation
$385 spent on food and drink
$150 spent on purchases
$150 spent on sightseeing
Each of these expenditure categories is taken through four multiplier rounds before a figure for the total national income generated is obtained. In the case of the $315 spent on accommodation the rounds look as follows:

Case study C (*continued*)

First round	Tourist pays to hotel.	$315.00
Second round	Hotel pays wages, taxes, and purchases local goods and services.	$248.37
Third round	Wage-earners buy goods and services, local suppliers pay wages, and purchase other goods and services, government spends tax revenue on local economy.	$151.91
Fourth round	Same as third round.	$ 61.00
Estimated total spending		$776.28

The total annual turnover per dollar of 2.46 is calculated by dividing $776.28 by $315.00, indicating that the original expenditure on accommodation 'turns over' 2.46 times in a year. The same procedure was followed for food and drink giving 2.2; the accommodation figure of 2.46 was also used for sightseeing; and the food and drink figure of 2.2 was applied to purchases. The weighted average of all these came to 2.3 and was considered to be the value of the tourist multiplier.

The critics of the Zinder Report showed, however, that this 'multiplier' was not really a multiplier at all but some hard-to-determine measure of transactions or 'dollars changing hands'. Zinder had made the mistake of summing successive *gross receipts* at each round of spending rather than using the successive *incomes received* by residents. When the multiplier was recalculated in this way the total spending figure for accommodation dropped from $776.28 to $428.17, resulting in a multiplier of only 1.36.

A number of other serious omissions were also noted by the critics, leading them to make the conclusion that the real tourist income multiplier for the Eastern Caribbean was unlikely to be more than 1.0 for any island, or less than half the Zinder calculation. The case for advising governments to spend scarce resources on encouraging more tourism had not been made effectively and, 'No proper cost:benefit analysis of the returns from investment in the encouragement of tourism [was] performed. The social consequences of a rapid expansion of tourism [were] totally ignored'.

Economic analysis

Economic impacts outweigh other considerations in most assessments of tourism development in the Third World. However, the extensive literature has tended to be much more specific about expected benefits than costs, and extends from the discussion of broad effects on the whole economy to detailed impacts of a single new project. The unifying feature is a focus on things capable of being measured and evaluated by the tools of economic analysis.

Some idea of the many factors governing tourism's economic impacts is shown in figure 4.1. They imply a range of economic costs and benefits far greater than those brought into the Tanzanian debate, and suggest another level of analysis which should complement the political economic arguments we have already noted. There is insufficient space here to do more than choose a small selection of economic benefits and costs to illustrate the depth of investigation involved.

The balance of payments

The search by most Third World countries for hard currency with which to pay for modern industrial imports like machinery and motor cars puts international tourism high on the list of development priorities. Tourist spending gives rise to inward and outward currency flows which are generally thought to balance out in favour of Third World destinations, although there is evidence to show that few studies really investigate what happens in sufficient depth. These fiscal effects are generally categorized as follows:

Primary effects arising out of currency inflows from foreign visitor expenditure in a host country and outflows coming from the spending abroad by residents. They are recorded in various ways by banks and businesses and are fairly easy to measure.

Secondary effects arise as the direct expenditure is gradually felt in other sectors of the economy. They are divided into three categories of direct secondary effects (such as travel agent's commissions); indirect secondary effects as the tourist service industry passes some of its earnings on to other businesses (as when an airline contracts a local company to supply on-board meals which, in turn, means importing some of the food by that company); and induced secondary effects relating to the wages of those employed producing tourist goods and services. A proportion of this income may be remitted abroad by foreign employees.

Tertiary effects are the currency flows which do not come from direct tourist expenditure and relate to things like investment opportunities stimulated by

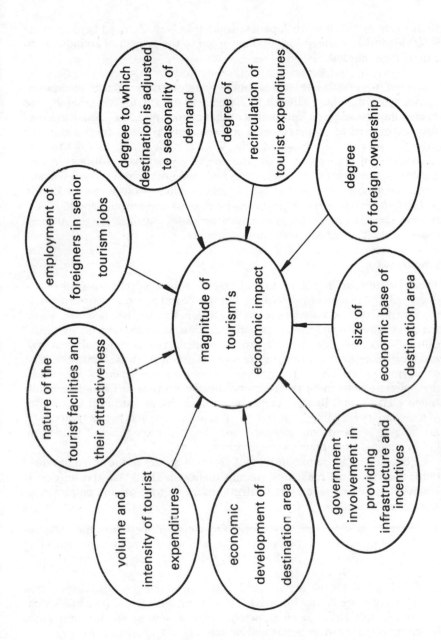

Figure 4.1 Factors governing tourism's economic impacts

tourist activity. Thus when Japanese tourists in New Zealand began buying large quantities of sheepskin products it led to the growth of an industry to export them abroad.

A failure to investigate the secondary and tertiary effects of tourism spending makes it very difficult to discover where it goes and what are the effects of its circulation. Net foreign exchange receipts are a good deal lower than gross earnings and reckoned to be as low as 50 to 60 per cent for countries in the Caribbean and the Pacific. One study found that the proportion of expatriates (non-nationals) in senior positions in the Caribbean tourist industry has been as high as 65 per cent (Cayman Islands) and accounted for over 48 per cent of all hotel and guest-house workers in the British Virgin Islands. A considerable but unknown amount of the income received by these workers was remitted to their homes abroad, reducing the net benefit to the host countries.

Employment

The effect of tourist expenditure on employment generation is also widely discussed in the economic literature but has been limited to answering only some of the chief questions involved. Little is known, for example, about the skills required and the returns and benefits expected; the geographical distribution of employment; the overall contribution to national, regional, and local economies; and the future significance of the travel industry as an employment generator. Three types of employment are generally recognized: first, direct employment from expenditure on tourism facilities like hotels (plate 4.1); second, indirect employment in businesses affected by tourism in a secondary way like local transport, handicrafts, and banks (plate 4.2); and finally, induced employment arising from the spending of money by local residents from their tourist incomes.

Once again a failure to consider all these categories has resulted in rather crude analysis of the real employment situation in Third World countries. It is possible to make several generalizations about tourism and employment from existing studies.

1 There is a close but not perfect correlation between the income-generating effects of tourism and the creation of employment, meaning that high returns from the industry do not correspond directly into proportionately more jobs.
2 The kind of tourist activity will influence employment because some types are more labour intensive. Jobs in restaurants and bars in Mexico, for example, have been shown to provide almost 40 per cent more employment than in hotels and motels.
3 The types of work skills available locally also have an effect on

Plate 4.1   Hotel employment in Sri Lanka

Plate 4.2   Handicrafts for sale, Colombo, Sri Lanka

employment, as illustrated by the high demand for unskilled labour. This has meant a preponderance of female employees in some Caribbean islands and the filling of senior jobs by expatriates.

4   Tourism may actually take employees away from other sectors of the economy or offer part-time jobs, both of which will have little effect in reducing overall unemployment figures.

5   Much tourism employment is seasonal; for example, in the Caribbean numbers of hotel employees per room vary from an 'in season' high of 1.4 to 0.9 'out of season'.

Some further and more detailed employment considerations are included in case study C.

Entrepreneurial activity

It is generally believed that tourism will develop backward linkages in an economy resulting in the creation of cost savings called external economies. This happens, for example, when an improvement to local services like transport or electricity is due to tourism but additionally provides a benefit to everyone in the area. The extent to which such improvements can actually promote business activity in a Third World country has not received much attention and the factors involved are more complex than might at first be supposed. Linkages between the hotel sector and local businesses will depend upon factors like the types of suppliers required (foodstuffs, maintenance and repairs, and so on); the capacity of local suppliers; the historical development of tourism in the area; and the type of tourist development.

When hotels are built gradually over time, the supply of local produce will keep pace with demand but may eventually outstrip capacity leading to a growing dependence on imported food. In the case of large new urban hotels which have been rapidly developed, there is a huge and immediate requirement for agricultural produce which can be met only from imports. It is hard for locals to break into this system of reliance on foreign suppliers and the dependence on imports remains. Unfortunately there is no tradition of providing the foodstuffs involved in many small island societies, in ensuring reliable supplies, or in seeking the entrepreneurial opportunities which make such businesses flourish elsewhere. An example of this problem can be seen in the two small Southern African countries of Swaziland and Lesotho where the Taiwanese government has developed demonstration farms. The vegetables produced are of show quality but are scarcely noticed by most local people who consider the whole exercise to be foreign to their own culture and of no application to themselves.

Economic costs of tourism

Several authors maintain that over-optimistic claims about the benefits of tourism twenty years ago have damaged the economies of some developing countries. Although this is probably true, modern feasibility studies are much more competent than the early examples (one of which is discussed in Case Study C), and it is possible to identify the main sources of economic cost.

Opportunity costs

The relative economic benefits to be gained from investing in tourism rather than some other industry is a comparison known in economics as the 'opportunity cost' of an investment. There are few studies which have been able to measure the values of the opportunities foregone as far as tourism is concerned. Young (1973) mentions the case of the Caribbean island of St Lucia where the coming of tourism resulted in a flight of labour from the local banana industry which was then the main source of foreign income. Those left on the land could not cope with the labour requirements of the banana crop with a consequent loss of productivity and earnings. Tourism led to a great increase in food imports as well and a great strain on the balance of payments. The net benefit to the island of the new industry was therefore marginal.

Over-dependence on tourism

Small Third World economies tend towards dependence on a single primary product and are badly affected by changes in commodity prices. In these circumstances the introduction of tourism appears initially as a welcome form of diversification. A problem arises (as in the case of St Lucia above) when tourism earnings supplant traditional activities and open the country's economy to instability of a different kind. Changes in tourism fashion, often dictated by transnational travel groups, can devastate such destinations and soon lead to a downward spiral of diminishing standards and poor publicity. This is all too evident in Jamaica where competition from new coastal resorts in Mexico's Cancun Peninsula is one factor which appears to have damaged a formerly buoyant industry.

As Young (1973) points out, the degree of acceptable reliance on tourism is also related to the structure of the economy:

> Where there is high unemployment, a relatively unskilled labour-force and few alternative sources of employment . . . then stimulation of the tourist industry may well be a correct course of action. The danger appears to be at the next stage of economic development, where unemployment and under-employment have been reduced, the labour-force is better

educated and an infrastructure exists which might support other industries. Continuing dependence and emphasis on tourism may no longer be economically justifiable.

Inflation

Although there are very few documented studies of how inflation due to tourism affects a host population, there is plenty of conventional wisdom to suggest that some price rises are linked to this cause. The more obvious instances involve increases in retail prices in shops during the tourist season, and steeply rising land values leading to a general rise in home costs and property taxes. Huge increases in land values in Swaziland's Ezulweni Valley tourist centre in the early 1970s are generally considered to have been due to pressure from South African investors and speculative buying from outside the country. The Swazi government countered this with a land speculation control act requiring all property to be first offered for sale to local buyers before permission could be granted by a review board for foreign purchase. The effect was to reduce land prices considerably and bring them back in reach of some local residents.

Key ideas

1   The economic effects of international tourism in the Third World is the subject of analysis at two broad levels. One is a wide-ranging and largely negative discussion which views tourism's economic consequences in the context of national development. The other is restricted to conventional tools of economic analysis, possesses a limited focus on economic issues, and is generally positive about the industry's prospects.

2   The primary economic benefits of tourism are generally regarded as: a contribution to foreign exchange earning and the balance of payments, the generation of employment and of income, the improvement of economic structures, and the encouragement of entrepreneurial activity. The evidence as far as Third World destinations are concerned is mixed.

3   The economic costs of international tourism are not well researched but are considered to include: increased inflation and land values, increased pressure to import, seasonality of production, problems connected with over-dependence on one product, unfavourable impact on the balance of payments, heavy infrastructure costs, and the effect on growth of having much of the labour-force employed in a service industry with poor productivity prospects.

4   Considerable care needs to be exercised in the use and interpretation of economic multipliers in tourism. The size of the multiplier varies according to the method used, the scale of the economy, the structure of the economy, the volume of imports used by tourists, and so on.

5
Tourism's environmental, social, and cultural impacts

International tourism results in a form of imported development with many physical and social repercussions in the Third World. The same effects may derive from other externally inspired changes in these countries but those due to tourism are sufficiently distinctive to have given rise to much environmental and social research. The question of wildlife conservation in Africa, for example, is a large subject in its own right but is linked to tourism because of fears that unrestrained poaching or poor management will do more than destroy the animals and their habitat but will threaten a lucrative industry as well. On the social side tourism leads to a 'revolution of rising expectations' and western consumerism. It also contributes to a push for modernization without the prior industrial phase of development experienced in Europe and North America in the last century.

The long-term interests of both rich industrialized and poor agrarian countries seem to coincide in the case of tourism's positive effects in conserving the environment but the same cannot be said of its primarily negative social impacts. Very few authors are impressed with the goodwill or promotion of understanding which is claimed to flow from international tourism, and prefer to concentrate on its more obvious problems. This chapter begins by looking at the different ways tourism reacts with the environment in a Third Word setting (figure 5.1) before considering an East African example of wildlife management in case study D. Extensive and recent social and cultural envidence is then examined in sections dealing with personal interactions between tourists and host society, the many factors contributing to social impacts and some recorded cultural effects. A

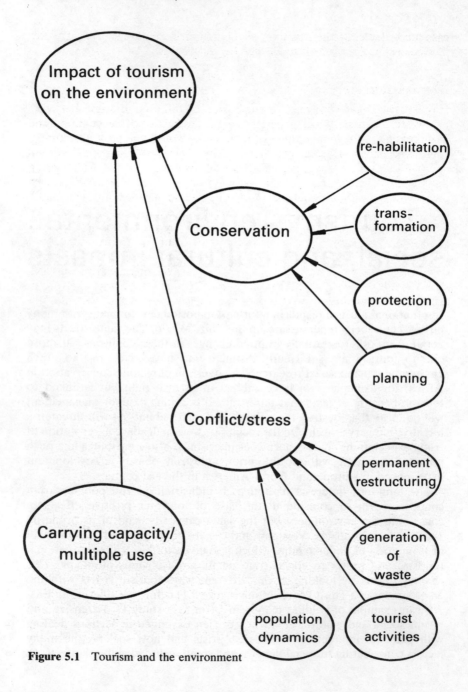

Figure 5.1   Tourism and the environment

case study looks at the sex tour phenomenon in south-east Asia and its particularly noticeable effects in the city of Bangkok.

Environmental impacts

It is necessary at the outset to re-emphasize the differing priorities between rich western nations, which generally see the destruction of the environment as a pressing problem affecting everyone in the world, and poor developing countries, whose priorities are first to raise their own living standards to acceptable levels. Many examples from all over the Third World suggest that much of the attraction for foreign visitors is an 'unspoilt' natural environment and traditional way of life of a kind long passed in Europe and North America. Such attractions, however, are also a sign of under-development and rarely tolerated just because they meet with the approval of visitors. One reason for visiting Singapore in years past, for example, was the existence of old Chinese shophouses and markets, but these were demolished in the building boom of the 1970s, leaving the government with an interesting problem. Whether to re-create Chinese tradition in new and artificial developments, giving the tourists a taste of what they expected to see, or to accept a change in image? The result has come down firmly on the side of modernization with the emphasis shifting to good accommodation, duty-free shopping, and entertainment in a new Asian setting (plate 5.1).

Plate 5.1 Luxury hotel in Singapore

Plate 5.2   Well maintained Georgian buildings in Spanish Town, Jamaica

Conservation influences

Although there is plenty of positive evidence about tourism's part in conserving and enhancing Europe's building treasures and historic sites, much less is written about this sort of impact in the Third World. Four main conservation influences are noted by Mathieson and Wall in their review of the worldwide literature, and these are listed here with some Third World examples.

First, there is the rehabilitation of existing buildings and historic sites. Notable progress has been made in some Asian countries, with the partial reconstruction and protection of ancient cities and temple shrines being encouraged because of their unique attractions for foreign tourists. Well known in this respect are Ayudhya, a ruined former capital of Thailand, the Buddhist temple at Borobudur in Java (Indonesia) and Anuradhapura, one of the ancient jungle cities in northern Sri Lanka; but this process is long and expensive and easily disrupted by signs of political and social instability. It is inadvisable today, for example, for foreigners to visit the famous temple at Ankor Wat in Kampuchea or the northern Sri Lankan historic sites because of these dangers, and conservation activity has ceased as a consequence. Fortunately in some countries care has been taken to restore and maintain a more recent architectural heritage as well, such as the Georgian buildings in Spanish Town, Jamaica, shown in plate 5.2.

Second, there is the transformation of old buildings to new uses. There is some evidence of this effect in India, for example, where the magnificent guest-house attached to the maharajah's palace in Mysore is now a luxury hotel. Third, there is the conservation of natural resources. This effect is seen widely in many parts of the Third World in the establishment and protection of national parks. Although examined in terms of wildlife and tourism later in the chapter, it needs to be emphasized here because the prospect of earnings from foreign visitors can provide the stimulus to go ahead with environmental conservation. This probably would not occur in most instances without tourism's promise of secondary benefits. The results in environmental terms are rather mixed, with evidence which now suggests that it is easy to exceed the tourist carrying capacity of prime attractions, with considerable damage resulting.

Finally, there is the introduction of planning procedures and controls to ensure good management of the environment. The popularity of some attractions demands management controls which might not have seemed necessary without tourism. A difficulty with this in the Third World is that such measures are expensive and possibly quite foreign to local customs and behaviour. Strict control over building permission, for example, is commonplace as a conservation device in the west but is difficult to enforce in a Third World setting where land may be held under communal (shared) tenure systems and not subject to legislative controls.

Tourism, conflict, and the environment

In 1977–8 the Organization for Economic Co-operation and Development (OECD) prepared a research framework for the study of tourism-induced environmental stress. Four main groups of stressor activities – changes causing permanent restructuring of the environment, the generation of waste products, tourist activities, and population effects – were identified according to the nature of the stresses involved and the environmental response. Although not developed specifically for the Third World, it provides a useful means of summarizing some of tourism's chief environmental conflicts.

Permanent environmental restructuring

This comes about as a result of major construction activity like a new highway or resort complex or even, as in the Maldives, a new airport. Besides the various physical effects on the natural environment, such developments result in the removal of large quantities of land from primary agricultural production or from its undeveloped status as a natural habitat. A chain reaction of environmental and human responses affecting things like changes in the populations of biological species and new demands on conservation expenditure soon follow.

Generation of waste products

Both the built environment and transport effects of tourism give rise to various residual products in the form of waste. The commonest of these are effects on water and air quality, neither of which seem to have been studied in any depth from a tourism perspective in the Third World. The effects of recreational activities on water quality and pollution have been reviewed by Wall and Wright, who list four main effects. First the pollution of water from raw or untreated sewage. This is a two-way consideration from a tourism perspective as some visitor developments cause pollution, whilst in others the tourists face a health hazard because of dirty water. The latter concern is endemic in the densely populated and tropical Third World. Second, there is the addition of nutrients to water which occurs as a result of effluent or run-off and leads to excessive weed growth and consequent reduction in water oxygen and fish levels. The third factor relates to the oil products from shipping and boating; these also affect oxygen levels adversely, as well as reducing the appeal of water for recreational purposes. Finally there are the petrol by-products from boating which introduce lead into the water with toxic effects on life forms. The presence of detergents and other domestic chemicals have very similar effects.

The quantitative effects of such tourism-inspired environmental stress are not known nor have they been studied as a priority in the Third World where there are few countries with the facilities to deal with pollution of any kind, let alone from tourism. However, this situation seems to be changing, and in 1985 a workshop was held on the environmental assessment of tourism development in the Republic of the Maldives. The important tourist industry of this Indian Ocean country depends on the unspoilt beauty of many small (less than 1 sq km) resort islands (plate 5.3) and it is this peculiar characteristic of the Maldivian resorts which has led to serious environmental stress. Fresh water from wells in the coral rock is easily polluted by sewage waste during the wet season. In 1982 an epidemic is thought to have originated in this way in Male (the capital) which resulted in an emergency call for more than twenty doctors from neighbouring Sri Lanka.

Tourist activities and the environment

There is a relatively large literature on the interrelationship between various tourist activities and the environment. This ranges from the effects of walking on coral reefs, to the destruction of beaches and parks caused by trail bikes and off-road vehicles. Once again, most research has been conducted in the developed parts of the world, although in this instance there is no reason to believe that major physical differences exist in a Third

Plate 5.3   Resort island in Male Atoll, Republic of the Maldives

World environment. An interesting example of ecological change resulting from tourism and conservation efforts comes from the southern (Pretorius Kop) section of South Africa's Kruger National Park. This area was first reserved as a game park in the 1920s after being subjected to many years of burning and cattle grazing by the small tribal African population. By the 1970s the effects of park enclosure had led to the growth of thick bush and a resulting change in the wildlife population from large savannah species to smaller and less easily seen forest animals. This is thought to have led to the unforeseen effect of reducing the popularity of this part of the park for visiting tourists because of their wish to see large game.

Tourist development and population dynamics

Tourism has a marked seasonal effect on population densities in all sorts of destination environments ranging from beach resorts to mountain areas and wildlife parks. In most of the tropical Third World such seasonality is governed by a combination of local climatic variations and the northern hemisphere holidays. This results in distinctive flows to warmer places in the south during the European and North American winter, as well as to an avoidance of tropical destinations during the months of greatest climatic stress. The most obvious of these is the hurricane season in the Caribbean and the equivalent cyclone season in the Pacific. Nature conservation itself through its removal of potentially good agricultural land may heighten population pressures elsewhere.

Case study D

Tourism-carrying capacity in an East African game park

The long-standing popularity of national game parks in Africa has given rise to many studies documenting the impact of tourism on wildlife. This ranges from adverse effects such as the disruption of animal feeding and breeding patterns to a general recognition that tourism has greatly assisted the process of nature conservation. At issue is the dilemma of how to cater for the tourists without causing lasting damage to the fragile ecology of the parks. Tourist minibuses, for example, have been observed approaching a hidden cheetah family which has served to attract vultures and lionesses, causing the cheetahs to abandon their kill. In Uganda, research has revealed that crocodile nests viewed by tourists are much more likely to suffer destruction from predator monitor lizards than those not visited by touring groups. Apparently the tourist boats cause female crocodiles to enter the water leaving their nests open to attack by lizards and baboons which then steal the eggs.

One East African park in particular, the Amboseli National Park in Kenya (figure D.1), has experienced severe problems of tourist congestion. Wesley Henry's research findings show that almost 80 per cent of tourists restricted their viewing to a small, 15 sq km, area located along the edge of the park's woodlands and swamps. At first it was thought that this pattern was a result of overall animal distributions and the concentration of game in this area during the dry season. More detailed investigation revealed that the wildlife is actually much less concentrated in the park than vehicle use tends to be (figure D.2), requiring the researchers to look for other reasons.

The most likely explanation was found to be the fact that only six animal groups account for more than 80 per cent of tourists' stationary viewing time in a park which contains 56 large mammal and 400 bird species. Just two animals, lions and cheetahs, accounted for more than half the tourists' time (figure D.3). Thus vehicles tended to be concentrated in the area where the probability of finding these predators was highest (plate D.1). As many as fifteen to thirty vehicles sometimes cluster around a single animal group marking a disturbing decrease in the park's amenity value. Recommendations to improve the situation include developing a road system to disperse tourists, new research to discover the long-term consequences of vehicle

Case study D (*continued*)

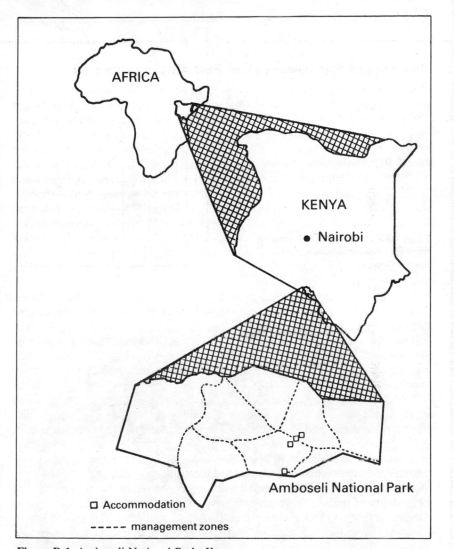

Figure D.1 Amboseli National Park, Kenya
Source: Henry (1980)

Case study D (*continued*)

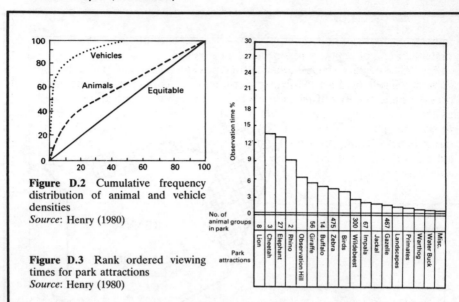

Figure D.2 Cumulative frequency distribution of animal and vehicle densities
Source: Henry (1980)

Figure D.3 Rank ordered viewing times for park attractions
Source: Henry (1980)

Plate D.1   Tourists viewing lion in Amboseli National Park

Case study D (*continued*)

impacts on the lion and cheetah populations, and efforts to gain a better understanding of what visitors desire and expect.

An often-neglected consequence of providing national parks is the difficulty faced by East African governments in coping with population pressures and the loss of tribal lands through wildlife conservation, as well as the sheer costs to the surrounding agricultural community of animal protection. The latter amount can easily be greater for local people than the direct benefits from tourism which has led some to suggest that controlled culling (or cash cropping) of the more abundant animal species should be considered by way of compensation.

Carrying capacity and multiple use

It is also important to identify the two planning concepts of carrying capacity and multiple use which enable host societies to make more efficient use of tourist facilities and land resources. Carrying capacity is a notion which recognizes that both natural and man-made attractions have upper limits in their capacity to absorb visitors, above which a deterioration of the resource itself takes place (see case study D). Although the concept is simple, its application is complex because of problems in measuring changes which occur, and discovering the causal relationships between tourist impacts and their effects on the environment.

Some idea of the complex interrelationships involved can be understood when it is realized that carrying capacity is affected by a range of tourist characteristics (their socio-economic profile, levels of usage, length of stay, type of activity, and levels of satisfaction), as well as by the various characteristics of the destination area and its population. Among the latter are environmental features (topography, flora and fauna, and so on), the level of economic development of the destination area, its social structure and organization, politics, and the scale of tourist development. All these things together will govern the capacity of a destination to absorb the demands placed on it by tourists.

Multiple use is a strategy which recognizes that a limited supply of recreational land often needs to be used for several purposes. It is possible, for example, for some forest recreation areas in North America to be used for logging as well as tourism. Pressures for similar combinations are most intense in small countries and islands where land is in very short supply. Such possibilities are clearly of interest to tourist islands in the Caribbean and Pacific, though it is difficult to find examples where multiple use strategies have been adopted deliberately.

Social and cultural impacts

There are three broad and complementary ways of examining the impact of international tourism on social conditions in Third World countries (figure 5.2). The first portrays the tourist–host encounter as an identifiable event with a number of positive and negative outcomes, the assessment of which depends on how the observer views the 'correct' path towards development. The second is a functional view of various elements of Third World society which may experience change as a direct result of tourism, such as moral behaviour, language, and health. The third perspective considers aspects of cultural change that come about through tourism's influence in resurrecting traditional skills and customs like handicrafts and dance.

Together, these three forms of social impact overlap a good deal and, as Mathieson and Wall point out, are hard to distinguish from one another in practice. The distinction between social and cultural studies is particularly hard to identify but is useful in differentiating between research into universal human issues like crime or health, and that directed at the things which condition human behaviour. In the latter instance certain products of this behaviour are found in the familiar artefacts of a traditional way of life in the Third World.

Tourist–host encounters

The sheer variety of tourist categories seen in the typologies discussed in chapter 3 ensures that tourist–host encounters will be equally varied and will depend on the stage of development of the tourist industry at a particular destination. Thus one can imagine some very positive outcomes from the infrequent contacts between 'explorer' tourists and local people in remote places, in contrast to the huge and often controversial impact of mass tourism elsewhere. According to a UNESCO report published in 1976, mass tourism gives rise to four main visitor–host relationships.

1   Transitory encounters are a feature of most temporary tourist visits and are viewed very differently on the part of traveller and host (figure 5.3). This was shown vividly in a scene from the Italian documentary film *Mondo Cane* (a dog's world) which portrayed North American tourists arriving to an organized welcome at Honolulu airport in Hawaii with flowers and a kiss from waiting hula girls. Close-up shots of the girls' faces before and after the obligatory kissing left the cinema audience in no doubt that this 'welcome' was sometimes an ordeal for the hosts.

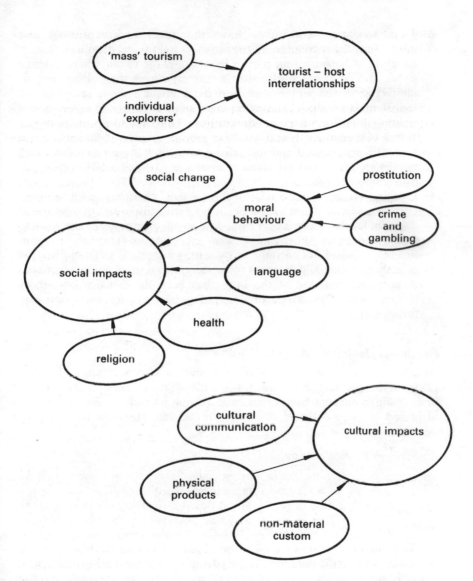

Figure 5.2 Social and cultural impacts of tourism

2   Time and space constraints have the effect of compressing and intensifying the encounter in order to meet tight travel deadlines. This is a feature of Japanese tour parties in Asia and the Pacific where a desire to see and do as much as possible scarcely allows the visitors time to interact with local people. Another characteristic is the development of tourist enclaves which can restrict encounters to hotel employees or the sighting of villagers from an airport bus as the only forms of contact.

3   A lack of spontaneity is also typical of most encounters, because they are generally pre-planned and formalized to fit in with tour schedules and usually involve a financial transaction such as payment to see a ceremony or a shopping expedition.

4   Unequal and unbalanced relationships usually characterize tourist–host encounters in the Third World because of wide disparities in wealth and different levels of satisfaction gained from meetings between the parties. The resentments thus caused are expressed in attempts to gain financially from brief encounters by setting a specially inflated price for tourists or demanding payment for undertaking normal public functions. A notorious example of the latter has been the practice by airport officers in parts of Africa of demanding payment for entry and exit formalities.

Functional identification of social impacts

Although there is wide agreement that tourism affects a range of social phenomena, it is much more difficult to attribute the extent of these impacts or to distinguish them from other modernizing influences. The categories identified below are therefore only a broad interpretation of a largely descriptive literature.

Tourism and social change

Attempts are often made in the social sciences to represent human behaviour over time as a series of stages in linear sequence. This often results in an oversimplification of reality and is descriptive rather than explanatory in purpose, but still manages to assist us in clarifying observations about change. In the case of tourism's social impact, one researcher has devised an irritation index of five stages of increasing disillusion to host society. Beginning with initial euphoria, then apathy, increasing irritation, outright antagonism and, finally, a stage when cherished values are forgotten and the environment destroyed by mass tourism. This model was based on research in Barbados and Niagara Falls but much of it is applicable to changing attitudes towards the industry throughout the Third World.

Hostility to tourism is summarized by the United Nations as a function of several underlying factors. The first of these relates to the physical presence

Figure 5.3 Transitory encounters

Source: *New Internationalist* Magazine (December 1984) (reproduced by permission)

of tourists. Effects on a host society will vary according to the size of the population in relation to the flow of visitors. This ranges from visitor levels equivalent to 1 to 2 per cent of the host population or less in large countries like Nigeria, to as much as five to ten times the number of local inhabitants in small island locations. In addition, there is a 'demonstration effect' which occurs when the presence of large numbers of tourists encourages consumption patterns which are inappropriate for the population as a whole. This can range from demands for expensive food imports to a newly found desire among the young to spend money on entertainment activities, like gambling, whose introduction often coincides with new tourist developments. Finally, there is the pervasive influence of neo-colonialism. As we have already indicated, international tourism follows the well-established trading and political connections of former colonial empires and is seen by some as evidence that imperialism persists. Frequently criticized is the foreign ownership of tourism plant (buildings and fixed developments) in the Third World and the filling of senior positions with non-nationals. These resentments are not confined to the tourist industry but seem to be exacerbated there because of ever-present reminders of foreign culture and former political and economic subjugation.

Case study E

Sex and tourism in Thailand

What a grand hotel it is: a white palace, delicately floodlit, fountains playing in the courtyard, trees festooned with white bulbs in the best possible taste. Though it's nearly midnight, eight lanes of traffic stand tangled, hooting and snarling, in the Bangkok street outside. Up the wide marble steps and in – through huge smoked-glass doors – to the foyer. You stop and blink. It's dark. You step forward, bump into someone, apologise, step forward again, bump into someone else. Then realise you are in a crowd. It's a well-dressed crowd (this is, after all, a grand hotel); their wallets bulge with *Baht*. Their eyes are bulging too. You follow their gaze.

They're looking at a huge shop window lit from the inside. It's the only light in the foyer. Behind the expanse of plate glass the goods are displayed on wide shelves that look like a shallow flight of stairs running the entire length of the shop. Deep-pile rose-coloured carpet, like velvet, covers the stairs, matched by folds of hanging drapes that clothe the walls. At this time of night business is good and the shelves are emptying fast. The goods are coded – by numbers on different coloured discs pinned to

Case study E (*continued*)

each one. You make your selection, pay the cashier, and – before you can pocket your change – your purchase is waiting to take you to your room.

What a bargain. Blue Number 33 has long shiny black hair cut into a thick straight fringe over eyes that are dark and slanting – but not too Chinky. Blue means body massage . . . for just 100 *Baht* [approximately £2.50 in 1984]. Tourists are bargain hunters, touring the global super-market, shopping for trophies – trinkets and triumphs – they could never afford back at home. Spain touts sunshine and sangria. Thailand specialises in sex.

This vivid scene written by Debbie Taylor for a special issue of the magazine *New Internationalist* describes sex tourism in Bangkok from the perspective

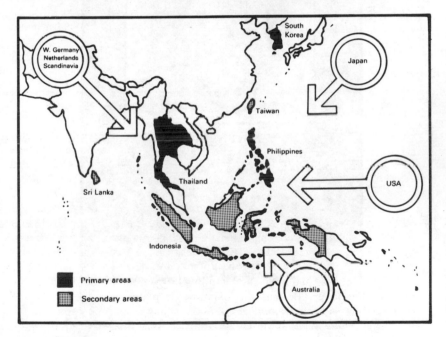

Figure E.1 Sex tourism in South-East Asia

Source: Seager, J. and Olsen, A. (1986) *Women in the World: An International Atlas*, London, Pluto Projects/Pan Books

Case study E (*continued*)

of the foreign visitors, many of whom holiday in Asia because of such attractions (figure E.1). In Thailand alone Cohen has noted more than half a million women reported as working in the sex industry, some 200,000 of

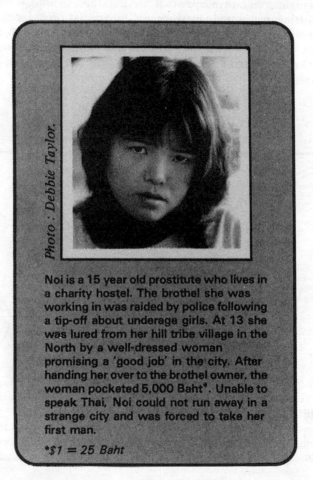

Noi is a 15 year old prostitute who lives in a charity hostel. The brothel she was working in was raided by police following a tip-off about underage girls. At 13 she was lured from her hill tribe village in the North by a well-dressed woman promising a 'good job' in the city. After handing her over to the brothel owner, the woman pocketed 5,000 Baht*. Unable to speak Thai, Noi could not run away in a strange city and was forced to take her first man.

*$1 = 25 Baht

Plate E.1 The story of Noi
Source: *New Internationalist* Magazine 142 (December 1984) (reproduced by permission)

Case study E (*continued*)

whom are prostitutes living in Bangkok. But what of the women themselves and the circumstances which have led to as many as 40 per cent of those working outside the home earning a living in this way? The story of Noi (plate E.1) seems to be typical.

The American anthropologist Nelson Graburn has reviewed the reasons why so many Thai and other Third World women are forced into prostitution and comes up with four main conclusions. First, a widespread patriarchal attitude which seems to cast women in either the role of madonna/virgin or whore; second, because women who have suffered seduction, rape, or been left by their male partners are often cut-off from other employment or marriage; third, many Third World countries are experiencing a crisis in their agricultural industry which forces rural people to search for employment in the cities; and fourth, women are discriminated against in most areas of formal employment and are left with the worst paid jobs.

Effects on moral behaviour

An extensive literature covering all the main tourist regions indicates that tourism is closely related to increases in the sale of sex (prostitution), crime of various kinds, and organized gambling. Although causal connections exist, it is very difficult to find hard evidence about the role played by tourism, given the fact that similar effects have been experienced in countries, like Nigeria, possessing low levels of tourism. It is thus wrong to make tourism the scapegoat for changes accompanying increasing modernization unless they are directly attributable to the industry.

In relation to prostitution, it seems that different and 'more relaxed' attitudes to sex in some Third World countries (as compared with the places where most tourists come from) were responsible for the growth of a sexual dimension to travel which dates back to the accounts of early European adventurers. This image has persisted in the marketing of some exotic destinations, like Tahiti or Thailand, and demands for such services now include almost everywhere affected by mass tourism (see case study E). The tiny West African country of The Gambia has been successfully marketed in Scandinavia as the closest place to Europe guaranteeing a pleasant climate during the northern winter. It is also a place deeply affected by tourism's impact on moral behaviour, as Harrell-Bond (1978) describes vividly: 'While female prostitution is common, male prostitution among young

Gambians is rampant. Middle-aged Scandinavian women, the age group which predominates among tourists, openly solicit'. She goes on to discuss how marriages between these partners sometimes eventuate with many couples returning to Scandinavia. For some of the young men the experience is a disaster: 'Their problems are so acute [they] have formed an association to attempt to give assistance of various kinds. There are stories of such Gambians who have been brought back to their country because they became insane'.

The very recent threat of AIDS as an incurable sexually transmitted disease will have important implications for the popularity of some tourist destinations, in addition to its major health effects on the entire populations of some regions like central Africa.

With respect to crime and gambling, although there is plenty of evidence in the developed metropolitan countries of a positive relationship between tourism and increases in crime, there are few studies which examine the situation from a Third World perspective. Mathieson and Wall have reviewed the literature and suggest that tourism and crime are influenced by population density during the tourist season; the location of a resort in relation to an international border (with considerable effects being seen on the USA–Mexican border); and differences in per capita incomes between host society and the tourists.

Health

Tourism has the dual effect of promoting the provision of improved health care in Third World destinations but, in addition, acts as a vehicle to spread some forms of disease. It is not uncommon to hear of tourists dying of relatively simple complaints in places with poor hospital facilities or where certain tropical diseases are endemic. Part of the problem is that western visitors occasionally disregard the fact that a low standard of public hygiene in parts of the Third World rarely allows for a safe water supply, and hospitals find it difficult to attract or retain the services of skilled medical personnel. It would be unreasonable to expect facilities for modern heart surgery in Tonga, for example, with a total population of about 100,000.

Effects on culture

The increasing presence of international tourists in the Third World is accepted by anthropologists as an important element in the process of acculturation, whereby people in contact borrow from each others' cultural heritage. In this instance the presence of 'stronger' western ideas and practices introduced by tourism means that the process is largely one of 'assimilation' of the 'weaker' host culture. This does not always happen, of course, as can be seen in those countries under powerful Islamic religious influence (like Malaysia and Indonesia) where active steps are taken to resist the forces of assimilation.

Mathieson and Wall have suggested a broad division of the extensive literature into studies investigating tourism's role in communication between cultures, its effects on the physical products of culture such as arts and crafts, and influences on custom and ceremony in host society.

Communication between cultures

Cross-cultural contact arising from tourism is thought to be a function of at least three factors. The first is the type of tourist. As suggested in chapter 3, the different categories of tourists in the Third World are reflected in expected differences in their kinds of interaction with local people. Second is the context in which the contact takes place. Clearly things such as length of stay, the environment under which the contact occurs, and language ability will help to determine the depth of communication which takes place. The frenetic pace and tight schedules characteristic of Japanese tours of Asia and the Pacific is an example of minimal cross-cultural communication, and has been studied specifically for major tourist destinations like Singapore. Finally, there is the role of cultural brokers, who are an intermediary occupational group such as interpreters and tour guides, who are the conduits through which much of the contact occurs. Their activities are thought to have a considerable effect on the manner and speed with which new ideas and influences are transmitted.

Tourism and physical culture

One of the most obvious signs of cultural reawakening or deterioration is to be found in the state of traditional art forms in Third World society. The growth of a tourist handicrafts market has stimulated local production in both positive and negative directions.

Positive influences may be found in the financial success of traditional art and artefact production in many places, but seems to have reached its greatest expression among the aboriginal populations in developed countries like Canada (the Eskimo or Inuit) and Australia (Aboriginal people). In Australia, for example, the ancient sand painting of desert tribes in the Northern Territory has been successfully adapted through the use of acrylic paints and canvas to form the basis of a lucrative export market where individual works can fetch tens of thousands of dollars. Further north, in Arnhem Land, 'bark paintings' have become a valuable commercial product with the use of modern glue to seal formerly fragile dyes.

On the other hand, the sheer pressure caused by a ready market for handicrafts has also lead to a fall in the quality of workmanship and the manufacture of cheap imitations known as 'airport art'. Some observers claim that traditional designs are degraded in this way and old skills lost and accusations abound of the existence of 'fake' art on sale from Africa to the Pacific. Most obvious is the production of small and unusable replicas of large items like the Australian *didgeridoo* (Aboriginal musical instrument), because the originals are far too big to be transported easily.

Tourism and local custom

Most readers will be familiar with the ability of modern tourism to market its products as a commodity, part of which consists of promoting the cultural attractions of holidaying in an exotic environment. Although this process is criticized for cheapening cultural events such as religious ceremonies, it is also responsible for the flow of funds into many local activities. Among the latter is the support of local musicians who might otherwise disappear under the onslaught of imported music spread through radio and television. In Jamaica, for example, foreign visitors are presented with a wealth of music from traditional dances and songs to reggae and western tunes. The survival of the former, the songs and dances from the plantation era, seems to be sustained to a considerable degree by the tourist entertainment industry.

But these events can go terribly wrong, as happened in Turner and Ash's account of an arts festival in Papua New Guinea in 1972 when one group of warriors, offended that they had not been awarded first prize, attacked the audience of tourists with bows and arrows.

Key ideas

1 Tourism has the secondary effects of conserving, as well as conflicting with, aspects of natural and man-made environments in the Third World.
2 Environmental conservation in very poor countries tends to be given a lower order of priority than raising living standards unless it coincides with income-earning possibilities such as the demands of the tourist industry.
3 The management concept of 'carrying capacity' provides a means by which the host society can establish thresholds to limit the numbers of tourists visiting attractions. The allied notion of 'multiple use' may enable a more efficient use to be made of scarce resources in small island communities.
4 Most informed opinion is negative about the social consequences of tourism's impact in the Third World, with particular criticism being levelled at the effects on moral behaviour.
5 There is considerable evidence to suggest levels of hostility and resentment in host societies towards tourism as the industry becomes increasingly institutionalized.
6 International tourism acts as a catalyst towards the assimilation of traditional custom by western culture in many Third World societies.

Exceptions to this generalization are those countries under Islamic religious influence.

7   Tourism has resulted in the commercialization of certain aspects of physical and non-material culture, such as the manufacture of artefacts for sale and the preservation of musical skills and ceremony.

6
Planning and management of tourism

The fact that few Third World governments have been successful in managing tourism for the overall benefit of their populations should come as no surprise in a situation where they have little control over the influences determining the size and character of the industry. This reality is not confined to tourism and applies to other sectors of world trade where many suppliers compete with each other for a share of a plentiful resource. As we noted in the introduction one warm place is very much like another when the product involved is made up of surf, sand, sun, and sex, and a wide gulf exists between what Third World governments would like to see in their domestic industry and what is realistically possible.

A major difficulty in tourism planning at the destination end of the business is to move from a list of hoped-for outcomes to a realistic agenda for action. More than a decade ago Robert Britton (1977) reported that the Caribbean island of St Vincent had prepared such a list as a means of building 'an indigenous and integrated' tourism industry comprising seven chief elements.

1 Zoning to separate tourism from other land uses and minimize its effect on agricultural land values.
2 Gradual growth to lessen inflation and social problems.
3 Indigenous tourism to maximize participation of local communities.
4 Local production of food, furniture, and crafts to stimulate the economy and save on foreign exchange.
5 Indigenous building forms using local materials plus more control over foreign investment.

6   Joint ventures between foreign and local investors plus more control over foreign investment.
7   Low-cost marketing aimed at reaching a diverse group of potential tourists.

Sadly the St Vincent proposals were frustrated by a change of government and opposition from large hotel groups. But even if this had not happened there are good reasons to doubt whether four at least of the suggestions could have been promoted by government without unforeseen effects on the island's share of the regional tourist market. Desirable objectives such as gradual growth, indigenous tourism, local production, and appropriate marketing, for example, may add up to little more than a signal to tour companies to shift their own promotion to other countries where there is less government intervention and the possibility, therefore, of greater profits. Little wonder then that tourism is often viewed as a last resort rather than development priority in places where few alternative means of earning foreign exchange exist.

In spite of this rather disappointing scenario it is possible to identify certain basic tasks which should be undertaken in all Third World countries with a tourism industry, as a means of adopting more appropriate policies. Robert Cleverdon, working for the British consulting group the Economist Intelligence Unit, has summarized the challenge by posing three central questions (figure 6.1):

1   What is the country's tourism resource potential?
2   How can the maximum benefit be obtained?
3   How is the country's tourist industry performing?

Examining each of the questions in turn enables us to address a selection of the issues involved, though it is not the role of an outside observer to prescribe which strategies to adopt. This task, as we have already stressed, belongs to Third World governments themselves even if such decisions are made in a global context which places little value on the exercise of such sovereignty.

Tourism potential

Most development specialists agree that tourism must be included in the plans and strategies for national growth and governments are cautioned to stop the industry expanding at a rate beyond local capabilities to control. Such an approach will help to minimize potential conflicts between competing land uses and demands that tourism resources are categorized according to their impact on the physical and human environment. In Fiji the *Eighth National Development Plan 1981–85* adopted four island

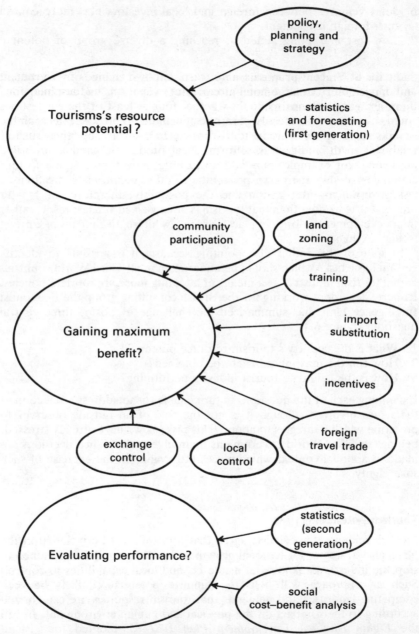

Figure 6.1   Three major questions in tourism administration (after Cleverdon 1979)

categories based on their capacity to accommodate different types of tourism:

1   Tourism resort islands which provide permanent overnight accommodation of international standard as the major economic activity.
2   Day visit islands which are uninhabited and tourist facilities are required to be related to environmental (carrying) capacity and designed for foreign and local visitors.
3   Local subsistence islands that are populated and depend mainly on agriculture and fishing. No major tourism to be encouraged though small-scale alternative forms, such as 'back-packing', are possible.
4   Island reserves that possess unique natural features which would be destroyed by development. Tourism is not permitted.

Such plans may be hard to implement in some Third World settings where much of the land, as in many African and Pacific nations, is held under customary (communal) forms of tenure and not subject to planning legislation. It is possible to draw up a simple table (table 6.1), however, suggesting where possible conflicts may occur.

Table 6.1   Fijian destinations according to type of tourism

Location	International	'Alternative'
Small islands		
tourist resorts	H	C
day visit islands	C	H
subsistence islands	C	H
island reserves	C	C
Main islands		
resort areas	H	C
native villages	C	H
scenic areas	C	H
reserves	C	C

Notes: C   indicates potential conflict
      H   indicates possible harmony of development
Source: Government of Fiji (1980)

Another factor minimizing the effectiveness of planning efforts is a general lack of standardized statistics on tourism measuring things like the length and location of visits and expenditure patterns. Guidelines defining these first generation data and various features of the tourism industry have been prepared by the United Nations and international comparisons are published regularly by the World Tourism Organization (wto).

Gaining maximum benefit

Community participation is often suggested as an essential ingredient in improving the quality of tourism's contribution to national development. It covers both the increased involvement of local people in decision-making about new developments, as well as various attempts at introducing something more adapted to local needs than the 'enclave model'. In this context, it must be remembered that increased participation actually means involving groups of low-income people in rural and urban areas who are not normally involved in the process of government. If popular participation is to be a reality it may require the introduction of new procedures in societies where such decisions are rarely made democratically. Among these 'alternative' tourism proposals is the Tourism of Discovery project pioneered in Senegal in the 1970s as a low budget exercise offering simple lodgings built, managed, and operated by the villagers. Visitors travel around a circuit of rest houses in small groups and have, according to reports, been well assimilated into the life of these communities without disruption.

Tourism training in the Third World is often criticized as being overly concerned with vocational and technical issues to do with western notions of running an efficient industry without paying enough attention to very basic concerns such as the cultural environment, communications difficulties with foreigners, and understanding of the rationale for the existence of tourism itself. When defined in these terms training takes on several new dimensions and is required to cover things like the non-verbal aspects of behaviour, aesthetic values, and concepts of comfort. David Blanton recounts, for example, that a hotel manager in Indonesia stressed that housekeeping training should not start with bed-making but with the concept of a bed!

Various financial incentives in the form of tax concessions or credit are commonly given by Third World governments to encourage investment in tourism, but there is little evidence to show how useful they are. Cleverdon suggests that it is easy to give too many concessions and that at the very least attention should be paid to:

1 The performance of incentive schemes in other countries.
2 The actual needs of potential investors.
3 Designing a code of concessions related to development objectives with requirements demanded of investors to be included (such as job creation targets).
4 Establishing performance standards capable of being monitored.

Third World countries can achieve greater control over their tourism industries but such measures all involve government intervention and none is very effective in situations where there are more beds than tourists to fill them. A general requirement is for *regional* groupings of countries to present a united front in negotiation with airlines and tour companies to increase their bargaining power. This is a good deal easier to advocate than achieve, however, and attempts at such regional co-operation in fields like transport have been unsuccessful in the past. Other regional possibilities such as marketing, data collection, and incentives have scarcely been explored in much of the Third World, but could have obvious benefits for the smallest countries.

Ultimately control over tourism means significant local ownership of the major sectors of the industry but the paradox remains that in the most vulnerable countries such investment may not be in the best interests of the community. In the words of an American planner commenting on tourism in Western Samoa:

> Maybe you don't want tourism or the hotel at all. You've built your hotel, but when you build a big hotel like this, you have to sell space at least one year in advance. You must let the market know about it. You didn't do that nor did you provide an airline service. If you really want to sell the hotel now you have to obtain regular airline service, perhaps enlarge the airfield, and possibly make other modifications. Maybe it's not worth it. Maybe you should consider whether your country really wants to bother with the tourism industry.
>
> (Farrell 1977: 123)

Evaluating performance

It is generally acknowledged that the evaluation of the tourism industry's performance is a task which has defeated most Third World governments. According to the Economist Intelligence Unit there are eight categories of second generation data which cover the minimum requirements:

1 Employment classified by age/sex; jobs; permanency; skill; location and expatriate/local status.
2 Population census material.
3 Consumer expenditure surveys.
4 Details of domestic savings.
5 Details of tourist industry costs according to items such as power, sewerage, rates, supplies, labour, and marketing. The identification of local against imported products is essential. Collected by annual surveys.
6 Balance of payments data.
7 An analysis of tourist expenditure from government department sources to cross check the survey in item 3 (above).
8 An analysis of government revenue including itemized customs duties and the sources of land taxes. This information can attribute the share of

imported goods received by hotels and provide a check on the answers given to the survey of tourist businesses.

Much of this information will already be available in the central statistical offices of some Third World countries but the sustained administrative effort in keeping it up-to-date is often underestimated. Co-operation is necessary among several ministries, requiring the generally lowly placed economic planning office to request data from much more powerful arms of government such as finance and the treasury (items 6 to 8).

Questions about the usefulness of some analytical methods, like the economic multiplier, in assessing tourism's impacts were raised in chapter 3. Social cost-benefit analysis is another approach often recommended in the literature as more effective because it covers the 'social justification for tourism development rather than simply the private profitability arising from it' (Cleverdon 1979). However, there are very few examples of its successful use in tourism planning because of the difficulties associated with measuring social, economic, and environmental effects in a uniform way. It seems irresponsible to promote the use of cost-benefit analysis in Third World tourism when so many questions about its effectiveness remain unanswered.

It will have become obvious by now that this introduction to tourism and development in the Third World is far from positive and stops well short of suggesting how the many imbalances and contradictions can be overcome. For most of us the question we might wish to ask is how to support more effectively those countries who are attempting to gain a better deal for their tourism industry. As the *New Internationalist* (1984) points out, however, we generally take a holiday to escape from reality, not to confront it:

> Maybe the healthiest thing we can do, as we lie on the beach on some distant shore, is to come to terms with what a tourist is – and to accept that we are using someone else's country for a kind of therapy. We should forget about the money we are providing. They are doing *us* a favour rather than the other way round. So if you feel uncomfortable as a tourist, if you feel ridiculous, if you feel a little bit humble, then you've probably got it just about right.

Key ideas

1 There is a world of difference between identifying hoped-for changes to the tourist industry and a realistic agenda for action.
2 An assessment of tourism industry potential should be included in the national development plan and requires the collection of standardized

data on visitor flows and the identification of suitable types of development.
3 Greater local control over the industry involves government intervention and may result in unacceptably high demands for investment in airports and the like at the expense of other national priorities.
4 Popular participation in tourism planning will involve special attention in societies where decisions about development are not usually democratic.

Review questions, references, and further reading

Items for further reading are indicated with an asterisk.

Chapter 1

1 What is it which distinguishes international tourism from other major sectors of world trade?
2 Try and identify those elements of tourism which contribute towards development.
3 What in your view are the main reasons why tourism is growing faster in some Third World regions than elsewhere (see figure 1.1)?

Cohen, E. (1974) 'Who is a tourist? A conceptual clarification', *Sociological Review* 22 (4): 527–55.
Friedmann, J. (1980) 'An alternative development?', in J. Friedmann, E. Wheelwright, and J. Connell, *Development Strategies in the Eighties*, Sydney, Development Studies Colloquium, University of Sydney, 4–11.
*O'Grady, R. (1981) *Third World Stopover*, Geneva, World Council of Churches.
*Pearce, D. (1987) *Tourism Today: A Geographical Analysis*, Harlow, Longman.
Senior, R. (1982) *The World Travel Market*, London, Euromonitor Publications.
Turner, L. (1976) 'The international division of leisure: tourism in the Third World', *World Development* 4 (3): 253–60.

*Turner, L. and Ash, J. (1975) *The Golden Hordes: International Tourism and the Pleasure Periphery*, London, Constable.

Chapter 2

1   What in your view are the chief strengths and weaknesses of the political economy approach towards international tourism?
2   Suggest ways in which generalizations about tourism in the Third World differ from descriptions of the industry in advanced industrialized countries like Britain and the USA.
3   Make a list and classify the different kinds of companies you were involved with as a traveller the last time you went away on holiday.

Britton, S.G. (1981) 'Tourism, dependency and development: a mode of analysis', *Occasional paper no. 23*, Canberra, Development Studies Centre, Australian National University.
*Crush, J.S. and Wellings, P.A. (1983) 'The Southern African pleasure periphery, 1966–83', *Journal of Modern African Studies* 21 [4] 673–98.
Lea, J.P. (1981) 'Changing approaches towards tourism in Africa: planning and research perspectives', *Journal of Contemporary African Studies* 1 (1): 19–40.
*Mathieson, A. and Wall, G. (1982) *Tourism: Economic, Physical and Social Impacts*, Harlow, Longman.
*United Nations Centre on Transnational Corporations (1980) *Transnational Corporations in International Tourism*, New York, United Nations.

Chapter 3

1   What are the arguments for and against creating images like the 'bliss formula' or the four *S*'s of 'surf, sand, sun, and sex' in Third World tourism marketing?
2   Can you think of any other ways of categorizing tourists visiting the Third World than the two typologies illustrated in this chapter?
3   If you had the choice of visiting any Third World tourist destination what kind of holiday would you choose and why?

*Cohen, E. (1979) 'Rethinking the sociology of tourism', *Annals of Tourism Research* 6 (1): 18–35.
Connell, J. (1987) 'Trouble in paradise: the perception of New Caledonia in the Australian press', *Australian Geographical Studies* 25 (2): 54–65.
Mathieson, A. and Wall, G. (1982) *Tourism: Economic, Physical and Social Impacts*, Harlow, Longman.
*Pearce, D. (1981) *Tourist Development*, Harlow, Longman.

*Smith, V. (1977) *Hosts and Guests: The Anthropology of Tourism*, Philadelphia, University of Pennsylvania Press.

Chapter 4

1   What in your view is the economic benefit from tourism most sought after by Third World governments? Find some examples to support your answer.
2   What benefit, if any, does international tourism bring to the living standards of ordinary people at holiday destinations in places like the Caribbean and Pacific islands?
3   Can you hazard a guess at what may happen to the future economy of a small Third World country mainly dependent on international tourism? Give some examples of trends and changes which have begun to take place in recent years.

*Bryden, J. (1973) *Tourism and Development: A Case Study of the Commonwealth Caribbean*, Cambridge, Cambridge University Press.
Bryden, J. and Faber, M. (1971) 'Multiplying the tourist multiplier', *Social and Economic Studies* 20 (1): 61–82.
Levitt, K. and Gulati, I. (1970) 'Income effect of tourist spending, mystification multiplied: a critical comment on the Zinder Report', *Social and Economic Studies* 19 (3): 326–43.
*Murphy, P. (1985) *Tourism: A Community Approach*, New York, Methuen.
Pearce, D. (1981) *Tourist Development*, Harlow, Longman.
Shivji, I. (ed.) (1973) *Tourism and Socialist Development*, Dar es Salaam, Tanzania Publishing House.
*Young, G. (1973) *Tourism: Blessing or Blight?* Harmondsworth, Penguin.

Chapter 5

1   What are the chief reasons for and against a poor Third World country acting to conserve natural treasures when such places may contain valuable resources such as oil or other commercial products?
2   Is it possible in your view to weigh the advantages from environmental conservation and a successful tourist industry against the disadvantages of negative social impacts of the kinds mentioned in this chapter? Can you think of ways in which such an assessment could be made?
3   See if you can identify the main difficulties in using a concept like 'carrying capacity' in the management of tourism development in a Third World country.

Cohen, E. (1982) 'Thai girls and Farang men: the edge of ambiguity', *Annals of Tourism Research* 9: 403–28.

*de Kadt, E. (ed.) (1979) *Tourism, Passport to Development?*, New York, Oxford University Press.

Graburn, N.H.H. (1983) 'Tourism and prostitution', *Annals of Tourism Research* 10: 437–43.

Harrell-Bond, B. (1978) 'A window on the outside world, tourism and development in the Gambia', *American Universities Field Staff Reports no. 19*, Hanover, New Hampshire.

Henry, W.R. (1980) 'Patterns of tourist use in Kenya's Amboseli National Park: implications for planning and management', in D. Hawkins, E. Shafer, and J. Rovelstad (eds) *Tourism Marketing and Management Issues*, Washington DC, George Washington University, 43–57.

*Mathieson, A. and Wall, G. (1982) *Tourism: Economic, Physical and Social Impacts*, Harlow, Longman (Chps 4 and 5).

*Organization for Economic Co-operation and Development (1980) *The Impact of Tourism on Development*, Paris, OECD.

*United Nations Economic, Social and Cultural Organization (1976) 'The effects of tourism on socio-cultural values', *Annals of Tourism Research* 4: 74–105.

Wall, G. and Wright, C. (1977) *The Environmental Impact of Outdoor Recreation*, University of Waterloo, Ontario, Department of Geography.

Chapter 6

1   What are the chief factors likely to prevent greater local control over the tourist industry in a small island nation like Fiji?
2   In what ways do you think the syllabus of a tourism training programme in the Third World might differ from that of a similar course offered in western countries?
3   Do you think the ideal of popular participation in tourism planning is possible or desirable in most Third World countries?

Blanton, D. (1981) 'Tourism training in developing countries: the social and cultural dimension', *Annals of Tourism Research* 8: 116–33.

*Britton, R.A. (1977) 'Making tourism more supportive of small state development', *Annals of Tourism Research* 4: 268–78.

*Britton, S. and Clark, W.C. (1987) *Ambiguous Alternative: Tourism in Small Developing Countries*, Suva, University of the South Pacific.

*Cleverdon, R. (1979) *The Economic and Social Impact of International Tourism on Developing Countries*, London, Economist Intelligence Unit.

Farrell, B.H. (ed.) (1977) *The Social and Economic Impact of Tourism on*

Pacific Communities, University of California Santa Cruz, Center for South Pacific Studies.

Government of Fiji (1980) *Fiji's Eighth Development Plan 1981–85*, vol. 2, Suva, Central Planning Office.

*Ritchie, J. and Goeldner, C. (eds) (1987) *Travel, Tourism and Hospitality Research: A Handbook for Managers*, New York, John Wiley.

Index

Page references in *italics* refer to maps or diagrams